How Science Made God Unnecessary

The Scientists & Discoveries That Replaced Divine Explanation

(Book 2)

Dr Prabhat Das

MD, PhD, FACP

No God Required

How Science Explains Reality Without the Divine

Hardcover ISBN: 979-8-951067-03-6

Paperback ISBN: 979-8-951067-02-9

eBook ISBN: 979-8-951067-01-2

Printed in the United States of America

TABLE OF CONTENTS:

Author's Note 1

Introduction: From God's World to a Lawful World 6

Chapter 1: When God Explained Everything 33

Chapter 2: The First Rule of Science: Look at Nature 59

Chapter 3: When Earth Lost the Center 89

Chapter 4: Galileo and the Fall of the Perfect Heavens 108

Chapter 5: Kepler and the Sky Written in Mathematics 129

Chapter 6: Newton and the Universe Under One Law 151

Chapter 7: Laplace and the Unneeded Hypothesis174

Chapter 8: When the Body Stopped Being Sacred Mystery 197

Chapter 9: Matter Without Magic 222

Chapter 10: Deep Time and the Death of Sacred Chronology... 246

Chapter 11: Darwin and Design Without a Designer 271

Chapter 12: Pasteur, Koch, and Disease Without Demons.. 295

Chapter 13: The Universe Beyond God's Old House 322

Chapter 14: Freud, Psychology, and the Human Need for God 348

Chapter 15: Neuroscience and the Natural Mind374

Conclusion: The Shrinking Space for God 398

Appendix: Scientists, Scholars, and Books That Changed the Human Understanding of Reality423

Thank You and Please Review the Book 449

Author's Note

Why I Wrote This Series

Some books begin in libraries. Some begin in classrooms. This series began much earlier, in the mind of a curious boy growing up in India.

I was born into a Hindu religious family. Religion was not a distant theory in my childhood. It was part of daily life. It was in prayers, rituals, festivals, stories, customs, family conversations, and the quiet assumptions that surrounded me. God was not discussed as a possibility. God was treated as reality.

My father was a firm believer in an omnipotent God. To him, God was real, powerful, protective, and deeply involved in human life. His belief was sincere. It was part of his strength, his discipline, and his view of the world. I grew up respecting that seriousness.

But from my teenage years, another force began to grow inside me. Curiosity.

I wanted to know why things happened. I wanted to understand nature, disease, death, the stars, the human body, and the mind. I was not satisfied with an answer simply because it was old, sacred, or repeated by many people. I wanted to know whether it was true.

In high school, I became deeply impressed by the lives and writings of great scientists. I still remember going to the local library and borrowing books written by, or about, Newton, Darwin, Einstein, and others. Those books were not popular among students at that time. In fact, I may have been the first student to borrow some of them. They were covered with dust. Some had even been damaged by insects.

But to me, they were treasures. Those dusty books opened a new world.

Newton showed me that nature follows laws.

Darwin showed me that life has a history far deeper than ancient stories.

Einstein showed me that reality is stranger, deeper, and more beautiful than ordinary imagination.

These thinkers did not ask me to believe blindly. They invited me to understand. They did not frighten me with punishment. They opened the door to wonder. They showed me that truth does not become weaker when questioned. It becomes stronger.

That invitation changed my life.

My curiosity did not end with school. It continued through medical college, scientific training, clinical practice, and decades of work as a physician. In medicine, I learned the value of evidence. A diagnosis cannot be accepted because it feels comforting. A treatment cannot be trusted because it is traditional. A belief about the body must face reality. It must face observation, testing, correction, and outcome.

That same habit of mind gradually shaped the way I looked at religion & God.

Even now, at retirement age, I continue to read almost every day. Science, philosophy, history, religion, psychology, evolution, cosmology, neuroscience, and the human mind still fascinate me. The more I read, the more I see how far human knowledge has come. Many things once explained by God are now understood through natural causes. Thunder, disease, eclipse, famine, fertility, mental illness, death, life, and the vast universe itself no longer need supernatural explanation.

This book series, "**No God Required: How Science Explains Reality Without the Divine", is a 7 book series**, and is my attempt to share that long journey in clear and simple language.

I do not write these books to insult believers. Many believers are kind, sincere, moral, generous, and deeply thoughtful people. I know this from my own family, my own culture, my patients, my friends, and my life experience. Faith often gives people comfort, hope, discipline, community, and emotional strength.

But faith does not decide truth.

A belief may comfort us and still need examination. A tradition may be ancient and still be mistaken. A prayer may bring peace and still not prove that anyone is listening. A sacred story may be beautiful and still be human-made.

My goal is to understand how belief began, why it became powerful, and why it still feels so real to so many people.

Why did human beings create gods?

Why did religion become one of the strongest forces in history?

Why do beliefs learned in childhood feel like truth?

Why do different cultures create different gods?

Why did science gradually make divine explanations unnecessary?

Can human beings live meaningful, moral, compassionate, and courageous lives without God?

These are serious questions. They deserve serious answers.

This series is based on knowledge gathered from many great books, scholars, scientists, historians, philosophers, psychologists, and thinkers. I have tried to bring their insights together in language that an ordinary educated reader can understand. My hope is that these books will encourage readers to think more rationally, question inherited beliefs, respect evidence, and study these subjects more deeply.

I also hope they will show that life without God does not have to be empty, cold, selfish, or immoral.

A life guided by reason, compassion, truth, science, justice, and human responsibility can be deeply meaningful. We do not need divine command to care for one another. We do not need fear of hell to be moral. We do not need supernatural stories to feel wonder before the universe.

Human beings are already capable of love, courage, kindness, beauty, and truth.

The universe is already astonishing.

Life is already precious.

Our responsibility to one another is already urgent.

No God is required for that.

Introduction: From God's World to a Lawful World

The Day the Sun Disappeared

Imagine a village long before science.

It is midday. People are working in the fields. Children are playing near their homes. Animals move lazily in the heat. The Sun stands high above the land, bright, familiar, and dependable.

Then something frightening begins to happen.

A dark bite appears at the edge of the Sun.

At first, people stare in confusion. Then the light begins to fade. Shadows sharpen. Birds grow quiet. The air cools. The ordinary world suddenly feels unsafe. Faces turn upward. Work stops. A strange silence spreads through the village.

The Sun, the great lamp of life, is disappearing.

No one in that village knows about the Moon's orbit. No one understands celestial mechanics. No one can calculate the path of an eclipse. No one knows that the darkness will pass and the Sun will return.

To them, the sky itself seems wounded.

Some fall to their knees. Some cry out to the gods. Some beat drums to frighten away an evil force. Some priests lift their hands toward heaven. Some whisper that the gods are angry. Some fear that a terrible punishment has begun.

And why should they not think that?

They have no telescope. No astronomy. No physics. No scientific calendar. No natural explanation.

They have fear, memory, imagination, ritual, and inherited stories. In such a world, a supernatural answer does not feel strange. It feels obvious.

Now imagine another scene.

A child is burning with fever. His mother holds him through the night. His breathing is fast. His skin is hot. His eyes are dull. The family gathers around him, helpless and terrified.

They do not know about bacteria. They do not know about viruses. They do not know about oxygen levels, antibiotics, vaccines, sterile technique, or intensive care.

What can they do?

They pray. They call the priest. They wonder if someone has sinned. They fear a curse. They ask whether a demon has entered the child's body. They promise offerings. They search their conscience. They beg God for mercy.

This is not stupidity.

It is human helplessness in the darkness of ignorance.

When people do not know the real cause of suffering, they still need an answer. If they cannot find a natural cause, they often imagine a supernatural one.

That is how God entered so many parts of human life.

God stood behind thunder.

God stood behind plague.

God stood behind harvest.

God stood behind childbirth.

God stood behind madness.

God stood behind kings.

God stood behind victory and defeat.

God stood behind death.

God stood behind the stars.

For most of human history, God was not merely a private belief. God was the great explanation that held the world together.

Book 1 of this series explained why human beings created God. It showed how God-belief grew from fear, death, nature, dreams, grief, family, ritual, tribe, authority, and culture. Human beings created gods not because they were foolish, but because they were frightened, intelligent, emotional, social, and deeply aware of suffering and death.

This book begins at the next great turning point.

What happened when human beings stopped only praying to the sky and began studying it?

What happened when they stopped seeing disease only as punishment and began looking for its physical cause?

What happened when they stopped asking, "Which god did this?" and began asking, "How does nature work?"

The answer is one of the most important stories in human history.

It is the story of God becoming unnecessary as an explanation.

The Question That Changed the World

The most powerful revolutions do not always begin with armies, kings, or battles. Sometimes they begin with a question.

What if nature can explain itself?

That question changed the world.

It did not appear everywhere at once. It did not remove religion overnight. It did not make human beings suddenly wise. But slowly, painfully, and often against resistance, a new way of thinking began to grow.

People began to observe before they concluded.

They began to measure before they believed.

They began to test before they trusted.

They began to correct old ideas when evidence proved them wrong.

For centuries, many answers had come from priests, scriptures, traditions, philosophers, kings, and inherited authority. The older world often asked, "What have we been taught?"

Science asked a more dangerous question:

"Is it true?"

That question was dangerous because it could overturn sacred certainty.

If the Earth moved, then human beings were not standing at the fixed center of creation.

If the Moon had mountains, then the heavens were not the perfect divine realm people had imagined.

If planets moved by mathematical law, then they did not need angels to push them.

If disease came from microbes, then fever was not a demon, a curse, or punishment from God.

If species changed through natural selection, then life could look designed without being directly designed.

If the mind depended on the brain, then even the soul began to look less certain.

Science changed more than knowledge.

It changed humanity's place in reality.

Three Meanings of God

Before going further, we must be clear about one important point.

The word God does not always mean the same thing in every argument.

For many ancient people, God meant a direct cause of events. God sent rain. God caused thunder. God punished through disease. God blessed harvests. God gave victory in war. God caused fertility, famine, plague, madness, and disaster. This was the God of immediate explanation.

For many believers, God also means a designer. God designed the human body. God designed the eye. God designed animals, plants, organs, instincts, and the order of nature. This is the God of design.

For more philosophical believers, God may mean something more abstract. God may be the first cause, the ground of existence, the source of natural law, the reason there is something rather than nothing, or the foundation behind reality itself. This is the God of ultimate explanation.

This book mainly examines the first two meanings.

It asks whether God is still needed to explain the natural world and its workings. Is God needed to explain the sky, planetary motion, anatomy, disease, chemistry, geology, evolution, the expanding universe, or the mind?

The answer, again and again, is no.

The third meaning of God, the God of first cause and ultimate reality, is important. It deserves serious discussion. That subject will become especially important in Book 3, where we will examine cosmology, the Big Bang, fine-tuning, first-cause arguments, and the claim that the universe requires a creator.

But Book 2 has a more focused purpose.

It asks what happened to God as a working explanation of nature.

And here the historical pattern is clear.

Where science found reliable natural causes, God lost explanatory territory.

How to Read This Book

This book is not an attack on religious people.

It is not saying that religious people are unintelligent, immoral, or incapable of doing science. That would be false and unfair. Many religious people have been thoughtful, compassionate, educated, and scientifically serious. Many great scientists in history believed in God.

Some even saw their scientific work as a way of understanding divine creation.

This book is also not saying that science has answered every possible question.

It has not.

Science does not remove grief. It does not make death easy. It does not promise perfect justice. It does not answer every moral question. It does not fully explain consciousness. It does not settle every philosophical debate about why there is something rather than nothing.

The question in this book is more precise.

When human beings used God to explain nature, did that explanation remain necessary after science found natural causes?

That is the question.

Does God still explain thunder better than atmospheric electricity?

Does God still explain infection better than microbes?

Does God still explain planetary motion better than gravity and mathematics?

Does God still explain fossils and rock layers better than geology?

Does God still explain life's diversity better than evolution?

Does God still explain memory, personality, and behavior better than neuroscience?

This book argues that the answer is no.

God may remain emotionally important to many people. God may remain culturally powerful. God may remain meaningful in prayer, ritual, grief, family, music, memory, and identity.

But emotional importance is not the same as explanatory necessity.

That distinction is central to everything that follows.

Science Did Not Defeat God by Insult

This book is not the story of angry atheists defeating religion.

That would be too simple. It would also be historically dishonest.

Many of the greatest scientists in history believed in God. Newton believed in God. Kepler saw mathematical order in religious terms. Boyle was deeply religious. Georges Lemaître, who helped develop the idea of an expanding universe, was also a Catholic priest. Many early investigators of nature did not think they were destroying religion. Some thought they were studying God's handiwork.

We should be honest about that.

But we must also be honest about something else.

A scientist's personal belief is not the same as the scientific meaning of his discovery.

Newton may have believed in God, but his laws made the heavens understandable without special divine intervention. Darwin did not write merely to attack religion, but natural selection made biological design possible without a designer. Pasteur was not writing a book against God, but germ theory helped remove demons, curses, impurity, and divine punishment from the explanation of disease.

This is the careful argument of this book:

Science did not make God unnecessary by ridicule.

Science did not make God unnecessary by hatred.

Science did not make God unnecessary by political ideology.

Science made God unnecessary by explaining reality better.

That is a stronger argument because it does not depend on mocking believers. It does not require us to pretend that all scientists were atheists. It does not force us to claim that science has answered every possible question.

It simply asks us to look at history.

Again and again, God was used to explain something mysterious. Then science found a natural explanation.

Once that happened, God was no longer needed for that explanation.

The eclipse no longer needed a dragon, demon, or angry deity.

The plague no longer needed a curse.

The orbit no longer needed a divine mover.

The mountain no longer needed a sacred myth.

The species no longer needed a separate miracle.

The fever no longer needed a moral judgment from heaven.

That is the pattern this book will follow.

The Old Answer and the New Explanation

The transformation from divine explanation to natural explanation can be seen clearly:

Old Question	Older Religious Answer	Scientific Explanation
Why does thunder shake the sky?	Divine anger or heavenly power	Atmospheric electricity and weather
Why does an eclipse darken the Sun?	Omen, curse, or warning from heaven	Motion of the Moon, Earth, and Sun

Old Question	Older Religious Answer	Scientific Explanation
Why do planets move?	Divine guidance or celestial spirits	Gravity, inertia, and mathematical law
Why do people become sick?	Sin, demons, curse, impurity, or fate	Germs, biology, environment, and physiology
Why does Earth contain fossils?	Sacred flood or divine mystery	Geology, deep time, extinction, and evolution
Why do living things look designed?	Divine designer	Natural selection and common descent
Why do religious experiences feel real?	Presence of God or spirits	Mind, culture, emotion, memory, and brain function

This table does not contain the whole argument. It only gives the reader a map.

The chapters that follow will tell the story in detail.

God Did Not Vanish. God Retreated.

God did not disappear in one blow.

There was no single day when science stood up and religion fell down. There was no one discovery that removed God from every human mind. There was no one book, one telescope, one microscope, or one equation that ended divine explanation.

Instead, something quieter happened.

God retreated step by step.

First from the sky. Then from the planets. Then from the anatomy of the body. Then from matter and chemistry. Then from the age of the Earth. Then from the origin and diversity of life. Then from disease. Then from the structure of the universe. Then from parts of the mind itself.

This retreat was not always admitted. It was not always welcomed. But it happened.

Before astronomy, the sky was full of divine messages. After astronomy, the sky became a physical system.

Before geology, Earth's history was often tied to sacred chronology. After geology, Earth became ancient beyond biblical imagination.

Before Darwin, life seemed to demand a designer. After Darwin, adaptation could be explained by natural selection.

Before germ theory, disease was often interpreted through sin, impurity, curses, demons, or divine will. After germ theory, disease became something that could be investigated, prevented, and treated.

This does not mean religion disappeared. It did not.

Religion remained in homes, temples, churches, mosques, scriptures, songs, rituals, memories, fears, and hopes. But something important changed. Religion lost more and more authority as an explanation of the natural world.

A belief may survive emotionally long after it has lost explanatory necessity.

That distinction is central to this book.

What "Unnecessary" Really Means

The title of this book is strong: *How Science Made God Unnecessary*.

But it must be understood correctly.

Unnecessary does not mean impossible.

Unnecessary does not mean every mystery has been solved.

Unnecessary does not mean science can comfort every grieving mother, answer every moral question, or remove the fear of death.

Unnecessary means something more precise:

God is not needed as a working explanation where natural causes explain the evidence.

We do not need God to explain eclipses. We do not need God to explain planetary motion. We do not need God to explain lightning. We do not need God to explain infection. We do not need God to explain why mountains rise and erode. We do not need God to explain why life changes over time. We do not need God to explain why Earth is not the center of the universe.

That is enough.

If God is no longer needed to explain the things God once explained, then the intellectual case for God becomes weaker.

God may remain emotionally meaningful to many people. God may remain culturally powerful. God may remain part of family, identity, music, ritual, and memory. But emotional importance is not the same as explanatory necessity.

A belief can comfort and still be false.

A belief can be ancient and still be mistaken.

A belief can be popular and still lack evidence.

A belief can feel sacred and still come from the human imagination.

This is not an attack on religious people.

It is an examination of religious explanation.

There is a difference.

Natural Explanation Proved Itself by Working

The great strength of science is not only that it gives a different answer.

It gives answers that work.

A predicted eclipse occurs at the predicted time.

A ship crosses the sea by using accurate astronomy and navigation.

A vaccine prevents disease.

An antiseptic operation saves a patient who might once have died of infection.

A blood test reveals what the eye cannot see.

A chemical drug relieves pain, kills bacteria, lowers blood pressure, or corrects a dangerous imbalance.

A telescope reveals worlds beyond ordinary sight.

A microscope reveals organisms invisible to the naked eye.

A satellite reaches orbit because mathematics, physics, and engineering are reliable.

This matters deeply.

A religious story may comfort. A ritual may unite a community. A prayer may reduce fear. But when we need to predict an eclipse, treat sepsis, build a bridge, sterilize an instrument, launch a spacecraft, or understand a fossil layer, natural explanation does the work.

Science proved itself by results.

This is why the movement from divine explanation to natural explanation was not merely philosophical. It was practical. It changed agriculture, medicine, navigation, public health, surgery, industry, communication, and daily life.

A correct explanation gives human beings power, not power over truth, but power to live more wisely within reality.

A false cause traps us.

A true cause opens a door.

That is why this history matters.

The Danger of Putting God in the Gaps

Human beings have often placed God in the gaps of knowledge.

If we do not know what causes thunder, God can be placed there.

If we do not know why a child dies, God can be placed there.

If we do not know why the planets move, God can be placed there.

If we do not know how life began or changed, God can be placed there.

If we do not understand consciousness, God can be placed there.

But a gap in knowledge is not evidence for God. A mystery is not proof. Ignorance is not a foundation for truth.

The wiser answer to mystery is not "God did it."

The wiser answer is, "We do not yet know."

That sentence is humble. It is honest. It leaves the door open to discovery.

"God did it" often closes the door too early.

This is one of the deepest differences between religion and science. Religion often gives a final answer before investigation is complete. Science accepts uncertainty and keeps looking.

That is why science moves forward.

It is not because scientists are morally superior. They are not. It is not because science never makes mistakes. It makes many. It is not because scientific institutions are free from pride, politics, ego, or error. They are not.

Science moves forward because it has a method for correction.

A scientific idea can be challenged. It can be tested. It can be improved. It can be rejected. It can be replaced.

Sacred certainty does not easily allow that.

Why This History Still Matters

This history is not only about ancient skies, old books, fossils, or laboratories. It matters now.

Even today, people still use God to explain disease, disaster, morality, politics, and personal suffering. Some still call illness punishment. Some still see natural disasters as divine judgment. Some still reject evolution because it threatens sacred belief. Some still trust miracles more than medicine. Some still turn to astrology, spiritual energy, magical healing, or supernatural signs when evidence would serve them better.

This is why the story remains urgent.

The battle is not simply between religion and atheism. It is between two ways of understanding reality.

One begins with inherited certainty.

The other begins with evidence.

One protects old answers.

The other allows correction.

One often turns mystery into doctrine.

The other turns mystery into investigation.

This matters because false explanations can cause real harm.

If disease is punishment, the sick may be blamed.

If mental illness is possession, the suffering person may be feared or abused.

If disaster is divine judgment, victims may be condemned instead of helped.

If evolution is denied, biology becomes harder to understand.

If prayer replaces treatment, preventable death may follow.

Science does not make life cold. In many cases, it makes life more humane.

It removes false fear.

It reduces false blame.

It gives us better tools to protect the vulnerable.

Wonder Without Superstition

Some people fear that science takes wonder away.

They imagine that if an eclipse is only the movement of celestial bodies, it is less beautiful. If life evolved, it is less precious. If the heart is a pump, love is less meaningful. If the brain produces thought, the mind is less profound. If the universe is not centered around us, human life is less important.

This fear is understandable.

But it is mistaken.

Understanding does not destroy wonder.

An eclipse is still magnificent when we know its cause.

A star is still beautiful when we know it is a sphere of hot plasma powered by nuclear fusion.

A flower is still delicate when we know it evolved through natural selection.

A child is still precious when we know the biology of development.

The brain is still astonishing when we know that memory, emotion, language, and imagination arise from neural activity.

Science does not empty the world.

It reveals what is actually there.

The universe does not need to be divine to be magnificent.

Life does not need to be designed to be precious.

The mind does not need to be a soul to be profound.

Morality does not need heaven to matter.

Human love does not need eternity to be real.

This is one of the deepest messages of this series.

A world without God is not a world without beauty.

It is a world without illusion.

What This Book Will Do

This book will follow the long historical movement from divine explanation to natural explanation.

We will begin in the older world, where God explained nearly everything. Then we will look at the first rule of science: look at nature.

We will enter the sky with Copernicus, Galileo, Kepler, Newton, and Laplace. We will see Earth lose its central position, the heavens lose their imagined perfection, and planetary motion become lawful and mathematical.

We will come down from the heavens into the human body. Anatomy and physiology will show us that the body is not a sacred fog of mystery, but a living structure with organs, vessels, tissues, circulation, and function.

We will examine matter and chemistry, where experiment replaced older ideas of magical transformation.

We will study geology, where Hutton and Lyell helped reveal deep time and an Earth far older than sacred chronologies had allowed.

We will meet Darwin, whose theory of natural selection showed how life could appear designed without being designed.

We will examine Pasteur, Koch, Lister, Semmelweis, and the rise of germ theory, where disease moved from demons and curses to microbes, sanitation, antisepsis, vaccination, and public health.

We will move into the modern universe with Einstein and Hubble, where space, time, gravity, galaxies, and cosmic expansion made the universe far greater than the small human-centered cosmos of earlier imagination.

Finally, we will turn inward. Freud, psychology, and neuroscience will help us ask why God feels so necessary and why even the mind may be understood as part of nature.

This book will not answer every question about the universe, life, morality, suffering, prayer, miracles, scripture, or atheism. Those deeper questions belong to later books in this series.

This book has one main purpose:

To show the historical pattern.

Where science explains, God retreats.

A Clearer World

This history is humbling.

It shows how often human beings have been wrong. We were wrong about Earth's central place. We were wrong about perfect heavens. We were wrong about the age of the Earth. We were wrong about the cause of many diseases. We were wrong about the separate creation of species. We were wrong about many things we once considered sacred and certain.

But this history is also liberating.

It means we are not trapped forever inside inherited errors. We can learn. We can correct ourselves. We can move from fear to understanding. We can replace superstition with knowledge. We can replace blame with treatment. We can replace fatalism with action.

This is one of science's greatest gifts.

Science does not promise emotional comfort in every sorrow. It does not pretend that life is easy. It does not remove death. It does not make the universe revolve around us.

But it gives us something precious.

It gives us respect for reality.

And respect for reality is the beginning of intellectual freedom.

The Question This Book Asks

The question at the heart of this book is simple:

What happens to God when nature begins to explain itself?

The answer will unfold chapter by chapter.

God does not vanish from human culture. God does not disappear from memory, music, ritual, family, or personal longing. God does not immediately leave the human heart.

But God becomes less necessary as an explanation.

That is the story of this book.

It is the story of the sky becoming astronomy.

The body becoming anatomy.

Disease becoming microbiology.

Earth becoming geology.

Life becoming evolution.

Matter becoming chemistry.

The universe becoming cosmology.

The mind becoming neuroscience.

It is the story of humanity moving from fear to inquiry, from sacred story to evidence, from inherited certainty to careful doubt, from divine explanation to natural explanation.

God was humanity's first answer to the unknown.

Science taught us that the first answer is not always the best answer.

Sources

The following sources helped shape the historical and scientific background of this Introduction:

Encyclopedia Britannica, "Scientific Revolution"

Stanford Encyclopedia of Philosophy, "Galileo Galilei"

Stanford Encyclopedia of Philosophy, "Newton's Philosophy"

Stanford Encyclopedia of Philosophy, "Religion and Science"

Encyclopedia Britannica, "Germ Theory"

Thomas Kuhn, *The Structure of Scientific Revolutions*

John Hedley Brooke, *Science and Religion: Some Historical Perspectives*

Peter Harrison, *The Territories of Science and Religion*

David Wootton, *The Invention of Science*

Jacob Bronowski, *The Ascent of Man*

Carl Sagan, *The Demon-Haunted World*

Chapter 1: When God Explained Everything

A World Full of Hidden Hands

Long before science, the world was not silent.

It spoke in thunder. It warned through eclipses. It punished through plague. It blessed through rain. It cursed through drought. It whispered through dreams. It shook the ground through earthquakes. It took children through fever and mothers through childbirth.

At least, that is how the world often seemed.

Imagine a farmer standing in a dry field, looking at a cracked sky that has given no rain for weeks. His crops are dying. His animals are weak. His children are hungry. He does not know about atmospheric pressure, ocean currents, climate patterns, soil moisture, or seasonal variation. He only knows that his family may not survive.

What should he think?

Perhaps the gods are angry.

Perhaps the village has sinned.

Perhaps an offering is needed.

Perhaps a priest must speak.

Perhaps someone has broken a sacred rule.

Now imagine a mother holding a child who is shaking with fever. The child cries, then becomes quiet. The family gathers around him with terror in their eyes. They do not know about bacteria, viruses, immune response, dehydration, pneumonia, meningitis, or sepsis.

What should they think?

Perhaps a demon has entered.

Perhaps God is testing them.

Perhaps a curse has fallen on the house.

Perhaps prayer is the only hope.

These were not foolish thoughts in the context of their time. They were human attempts to survive fear.

Before science, nature was full of hidden hands. Events did not merely happen. They seemed to be caused by someone or something unseen. The world felt personal, moral, and intentional. A storm was not only a storm. A fever was not only a fever. A famine was not only crop failure. Each event seemed to carry meaning.

Human beings stood in the middle of danger and asked the most urgent question of all:

Why?

For most of history, God answered that question.

Before Nature Had Natural Causes

There was a time when almost everything important in life seemed to come from above, below, or beyond.

Rain fell from the sky. Drought burned the fields. Rivers flooded without mercy. Lightning split trees. Earthquakes shook the ground. Children died of fever. Women died in childbirth. Crops failed. Armies won and lost. Kings rose and fell. Dreams brought back the dead. The night sky glittered with strange lights no human hand could touch.

Human beings were intelligent, emotional, frightened, hopeful, and curious. But they did not yet have modern science. They did not know about microbes, plate tectonics, atmospheric electricity, gravity, genetics, brain chemistry, or cosmic motion. They had eyes, ears, memory, imagination, family stories, tribal wisdom, rituals, priests, shamans, kings, myths, and sacred traditions.

So they did what human beings naturally do.

They searched for meaning.

If thunder shook the sky, something powerful must be speaking. If lightning struck a tree, some hidden force must be angry. If a child died suddenly, someone must have sinned, offended a spirit, or been cursed. If crops failed, perhaps the gods had withdrawn their blessing. If an eclipse darkened the Sun, perhaps heaven itself was warning the people.

Today, we call many of these explanations supernatural. But to the ancient mind, they did not feel supernatural. They felt natural. They were simply part of the way the world worked.

Before science, God explained everything.

Not because ancient people were stupid. They were not. They built cities, temples, irrigation systems, calendars, ships, tools, poems, laws, and civilizations. They observed the seasons. They studied the stars. They learned the habits of animals and plants. They developed medicine, architecture, agriculture, and moral systems.

But they did not yet have a tested, reliable, self-correcting method for explaining nature through natural causes.

Where knowledge was weak, imagination became strong.

Where evidence was missing, story filled the empty space.

Where fear was unbearable, religion gave comfort.

Where life felt chaotic, God gave order.

This was the old world. It was not a world without intelligence. It was a world without modern explanation.

The Human Need for an Answer

Human beings do not easily tolerate mystery when mystery threatens life.

A mystery in the stars may inspire wonder. A mystery in a child's fever creates panic. A mystery in a poem may delight us. A mystery in a spreading plague terrifies us. When suffering enters the home, people do not calmly say, "We will wait centuries for a scientific answer." They need meaning now.

Why did this happen?

Why now?

Why us?

Who caused it?

What does it mean?

How can we stop it?

What must we do?

Before science, religion answered these questions.

It gave causes. It gave rituals. It gave commands. It gave comfort. It gave someone to blame, someone to please, someone to fear, and someone to beg for mercy.

If rain did not come, people could pray, sacrifice, repent, dance, fast, or appeal to a priest. If disease struck, they could search for sin, impurity, demonic force, or divine anger. If an army won a battle, the gods had favored them. If an army lost, the gods had abandoned them. If a king ruled, perhaps heaven had chosen him. If disaster came, perhaps the people had failed their divine duty.

These explanations gave people something very important: a sense of control.

It may not have been real control. But emotionally, it mattered.

To say “the gods are angry” is frightening, but it also gives a path. Apologize to the gods. Offer sacrifice. Change behavior. Obey the priest. Follow the ritual. Purify the community. Find the offender. Remove the curse.

To say “we do not know” is honest, but it gives no immediate comfort.

That is why supernatural explanations became so powerful. They did not merely explain nature. They helped human beings live with fear.

Science would come much later and say, “Let us first find the real cause.”

Religion came earlier and said, “There is a reason, and the reason is divine.”

For frightened human beings, that answer often felt necessary.

The Sky as a Sacred Ceiling

For ancient people, the sky was not empty space. It was a living theater of power.

The Sun rose and set with majestic regularity. The Moon changed shape. The stars returned in patterns. Comets appeared without warning. Eclipses darkened the day or turned the Moon red. Thunder rolled across the heavens. Lightning flashed like a weapon. Rain fell from above and made life possible.

It is not hard to understand why the sky became sacred.

The sky was high, unreachable, brilliant, dangerous, and mysterious. It looked like the home of gods. The most powerful forces seemed to come from there. Light came from above. Rain came from above. Storms came from above. The seasons were marked by the movements of heavenly bodies.

So human beings gave the sky personality.

The Sun became a god or the eye of a god. The Moon became a goddess, a spirit, or a sign. The planets became wandering powers. The stars became ancestors, divine lamps, or messages. Thunder became the voice of a god. Lightning became a weapon. Rain became a blessing. Drought became punishment.

In many cultures, the sky was not only observed. It was interpreted.

People did not merely ask, "What is happening?"

They asked, "What does it mean?"

A comet could mean the death of a king. An eclipse could mean divine anger. A strange star could mean a birth, a disaster, a warning, or a change in power. The flight of birds, the shape of clouds, the direction of the wind, and the movement of celestial bodies could all be read as signs.

This is very important.

Before science, the universe was not mainly seen as a system of natural causes. It was seen as a system of messages.

Nature spoke. The gods warned. Heaven judged. The world was full of hidden meaning.

Science would later change this. Science would not ask whether lightning was a sign of anger. It would ask what lightning is. Science would not ask whether an eclipse was a warning. It would ask which bodies moved in front of which bodies, at what time, and according to what pattern.

That change sounds simple now. But it required a revolution in the human mind.

The older mind asked, “What is the divine meaning?”

The scientific mind asked, “What is the natural cause?”

The difference between those two questions changed civilization.

Disease as Punishment, Curse, or Possession

As a physician, I find the history of disease especially revealing.

Nothing shows the weakness of supernatural explanation more clearly than illness.

Before germ theory, before microscopes, before antibiotics, before vaccines, before anesthesia, before antiseptic surgery, people were surrounded by disease and death. A small wound could become fatal. A fever could kill a child in days. A woman could die in childbirth. A cough could spread through a household. A plague could empty a village.

People watched bodies fail without understanding why.

They could see the symptoms, but not the causes. They saw fever, rash, cough, swelling, diarrhea, pain, bleeding, delirium, paralysis, and death. But they could not see bacteria. They did not understand viruses. They did not know how infection spread. They did not understand immunity. They did not know why clean water mattered. They did not know why handwashing could save lives.

So disease became moral and spiritual.

It was explained as divine punishment, demonic possession, impurity, curse, fate, bad air, imbalance, or the will of God.

This was emotionally understandable. When disease strikes, people desperately want meaning. A random infection feels cruel. A divine punishment, however frightening, at least seems to have a reason. If sickness is punishment, perhaps behavior can change. If sickness is caused by a demon, perhaps the demon can be driven out. If sickness is a curse, perhaps the curse can be broken. If sickness is divine testing, perhaps endurance has spiritual value.

But there was a terrible cost.

When disease is misunderstood, the sick may be blamed. The patient becomes sinner, impure person, cursed person, possessed person, or moral warning. Instead of being treated, the sick may be feared. Instead of being studied, disease may be ritualized. Instead of looking for causes, communities may look for guilt.

Science changed this profoundly.

Once disease could be understood through natural causes, human beings could respond with treatment instead of blame. Infection could be prevented. Water could be cleaned. Hands could be washed. Instruments could be sterilized. Vaccines could be developed. Antibiotics could be used. Public health could save millions.

That transformation did not merely change medicine. It changed morality.

A sick child was no longer evidence of divine anger. A woman dying after childbirth was no longer proof of fate. An epidemic was no longer simply a punishment from heaven. Disease became a natural process that could be studied, prevented, and treated.

This is one of the clearest examples of the central argument of this book.

When the real cause was discovered, the supernatural explanation became unnecessary.

Disaster and Divine Anger

Natural disasters have always invited religious explanation.

An earthquake shakes the ground. A flood destroys homes. A volcano buries a town. A storm tears apart a village. A famine spreads hunger across a region. A fire consumes a city. A tsunami erases a coastline.

To people without geology, meteorology, seismology, oceanography, or climate science, such events seemed like acts of will.

The earth did not simply move. It was angry.

The river did not simply rise. It was punishing.

The volcano did not simply erupt. A god was speaking.

The storm did not simply form. Heaven was judging.

The famine was not only environmental failure. It was divine displeasure.

Again, this was not irrational within the limits of their knowledge. Human beings naturally look for agency. When something powerful happens, we ask who did it. This habit is useful in many ordinary situations. If a branch breaks near us, we look for an animal, a person, or a threat. If we hear a sound at night, we imagine an agent. If harm occurs, we search for intention.

But nature often has no intention.

An earthquake does not know who lives above the fault line. A hurricane does not choose the righteous and spare the wicked. A virus does not ask whether the patient prayed. A drought does not examine moral character. A volcano does not erupt because children failed to obey their parents.

Natural forces are not moral agents.

That idea is difficult, but necessary.

The older religious view often moralized nature. It turned disaster into judgment. Science removed that moral fiction. It showed that disasters have natural causes. Earthquakes come from movements of the Earth's crust. Storms come from atmospheric conditions. Floods come from rainfall, rivers, sea levels, land use, and geography. Famines come from weather, crop failure, politics, poverty, war, and distribution of food.

This does not make disasters less painful. But it makes them more understandable.

And when disasters are understandable, prevention becomes possible.

Buildings can be designed to withstand earthquakes. Weather systems can be tracked. Flood zones can be mapped. Vaccines can prevent epidemics after disasters. Public health systems can reduce suffering. Food systems can be improved.

Science does not stop all tragedy. But it gives human beings tools that prayer alone never gave.

Kings, Priests, and Sacred Authority

God did not explain only nature. God also explained power.

In many societies, kings ruled by divine approval. Priests spoke for gods. Laws were given sacred authority. Social order was placed under divine protection. To disobey the ruler could become more than a political act. It could become an offense against heaven.

This connection between religion and power is ancient and powerful.

A ruler who claims divine approval gains more than military strength. He gains moral authority. He becomes part of the sacred order. His commands become harder to question. His rule appears natural, cosmic, or God-given.

Priests also held great power because they stood between people and invisible forces. They interpreted omens. They performed rituals. They explained disaster. They declared purity and impurity. They forgave, condemned, blessed, or warned. They guarded sacred stories. They preserved tradition. They often controlled access to the divine.

Before science, this made sense. If the gods controlled harvest, disease, fertility, rain, war, and death, then the people who claimed to understand the gods naturally became powerful.

Religion gave structure to society. It united groups. It created shared rituals. It marked births, marriages, deaths, seasons, victories, and crises. It helped people cooperate. It gave moral rules. It gave identity.

But it also protected authority from questioning.

If a belief is sacred, doubt becomes dangerous. If a ruler is chosen by God, rebellion becomes sin. If a priest speaks for heaven, disagreement becomes disobedience. If a social order is divinely arranged, injustice can appear holy.

Science slowly weakened this structure.

Not by creating democracy directly. Not by solving all political problems. But by teaching a different habit of mind.

Evidence matters. Authority can be wrong. Tradition can be mistaken. Claims must be tested. Nature does not obey priests because priests are respected. The planets do not change their motion because kings command them. Disease does not spare the powerful because they claim divine favor.

The scientific way of thinking helped separate truth from authority.

That separation is one of the most important developments in human history.

The Moral Universe Before Science

Ancient people often lived in a morally charged universe.

The world did not simply happen. It responded. Good behavior might bring blessing. Bad behavior might bring punishment. Ritual purity might protect the community. Sin might bring disaster. Sacrifice might restore balance. Prayer might change fate.

This view gave meaning to suffering, but it also created confusion.

If a child died, someone might ask what sin caused it. If a city fell, people might ask which god was offended. If a woman was infertile, she might be blamed. If a person had epilepsy, psychosis, or depression, the condition might be interpreted as possession, curse, or moral failure.

The moral universe comforted some people, but it harmed others.

It gave answers, but often the wrong answers.

Science gradually separated natural events from moral judgment. It did not deny morality. It did not say good and evil are meaningless. It simply showed that many events once interpreted morally were not moral events at all.

Lightning is not a judgment.

A fever is not a sermon.

An eclipse is not a warning.

An earthquake is not a verdict.

A birth defect is not punishment.

A failed harvest is not proof of sin.

This separation is morally important.

When suffering is no longer seen as divine punishment, compassion becomes easier. When illness is no longer seen as impurity, treatment becomes more humane.

When disaster is no longer seen as God's judgment, victims can be helped instead of blamed.

Science did not remove morality from life.

It removed false morality from nature.

That was a great achievement.

Why Divine Explanation Felt Complete

Religious explanations were powerful because they answered many questions at once.

Science usually gives a narrower answer. It says how something happens. It identifies a cause, process, mechanism, or pattern.

Religion gives a larger emotional answer. It says why it happened, who caused it, what it means, what we should do, and how we should feel.

This is why religion was so satisfying before science.

If thunder is caused by atmospheric electricity, that explains the physical event. But it does not tell a frightened person whether the universe cares.

If infection is caused by bacteria, that explains the disease. But it does not tell a grieving parent why this child died.

If an eclipse is caused by the Moon passing between Earth and Sun, that explains the darkness. But it does not give the village a ritual, a story, or a moral lesson.

Science explains mechanism. Religion often combines mechanism, meaning, comfort, and command into one story.

That is why religious explanations can survive even after scientific explanations arrive. People do not only want facts. They want comfort, purpose, forgiveness, hope, and belonging.

This book does not deny that.

But emotional satisfaction is not the same as truth.

A comforting explanation can still be wrong. A meaningful story can still be mistaken. A sacred tradition can still misunderstand nature. A belief can help people endure life without accurately explaining the world.

We must be mature enough to see both sides.

Religion was useful in many ways. It gave community, ritual, discipline, courage, and comfort. But as an explanation of nature, it repeatedly failed.

Science was less emotionally satisfying at first. It did not say the gods were angry or pleased. It did not promise that every event had a moral meaning. It did not tell people that the universe was centered around them.

But science had one great advantage.

It worked.

The World Before Testing

The old world had many explanations, but few reliable ways to test them.

If a sacrifice was followed by rain, people could say the gods accepted the sacrifice. If rain did not come, they could say the sacrifice was insufficient, the ritual was impure, or the gods had a different plan.

If a sick person recovered, people could say God healed him. If the sick person died, they could say God chose another path.

If a king won a battle, the gods favored him. If he lost, the gods punished him, tested him, or transferred favor elsewhere.

This is the problem with explanations that cannot be tested.

They can survive every outcome.

A good explanation should be vulnerable to evidence. It should risk being wrong. It should allow reality to correct it.

Science became powerful because it accepted this risk.

If a prediction fails, the theory must be questioned. If an experiment cannot be repeated, the claim becomes weaker. If evidence contradicts the idea, the idea must change. If a better explanation appears, the older one must yield.

This is not easy. Scientists are human. They can be proud, stubborn, biased, ambitious, and wrong. Scientific communities can resist new ideas. But the method of science contains a corrective force. Evidence can eventually defeat authority.

Religious explanation usually does not work this way.

A sacred claim is protected because it is sacred. It is not easily surrendered to evidence. It may be reinterpreted, defended, spiritualized, or moved to another realm, but it is rarely abandoned by the same method that a failed scientific theory can be abandoned.

This difference matters.

Science can be wrong and still improve.

Religion can be wrong and remain certain.

The Beginning of a New Question

The old world asked, "What do the gods want?"

Science asked, "What does nature do?"

That shift changed everything.

It did not make human beings less emotional. It did not remove grief. It did not end death. It did not erase wonder. It did not answer every question.

But it gave humanity a new path.

Instead of fearing eclipses, people could predict them.

Instead of blaming demons for disease, people could search for causes.

Instead of reading comets as omens, people could study orbits.

Instead of seeing storms as divine anger, people could understand weather.

Instead of treating the body as a mysterious vessel of spirits, people could study organs, tissues, blood, and nerves.

This was the beginning of God's retreat as an explanation.

Not the retreat of God from human emotion.

Not the retreat of God from culture.

Not the retreat of God from poetry, ritual, or memory.

But the retreat of God from the serious explanation of nature.

That retreat did not happen in one century. It did not happen evenly. It did not happen without resistance. But once the habit of natural explanation began, it could not easily be stopped.

Every successful natural explanation raised a dangerous question:

If this can be explained without God, what else can be explained without God?

That question is the engine of this book.

The Strongest Religious Reply

A thoughtful believer may respond this way:

"Yes, people once used God to explain many things they did not understand. Yes, science has explained eclipses, germs, planets, storms, and many natural processes. But this does not disprove God. Perhaps God works through natural laws. Perhaps God created the universe with order. Perhaps science explains the mechanism, while religion explains the deeper meaning."

This is the strongest religious reply, and it deserves respect.

It is true that science does not disprove every possible idea of God. It is true that many religious thinkers have accepted natural laws. It is true that a person may believe that God works through nature rather than constantly interrupting it. It is also true that science is not designed to answer every question about meaning, value, or purpose.

But this reply also concedes something important.

It concedes that God is no longer needed to explain the natural event itself.

If lightning works through atmospheric electricity, we do not need divine anger to explain it.

If infection works through microbes, we do not need demons to explain it.

If planets move through gravity and inertia, we do not need angels to push them.

If species change through natural selection, we do not need separate acts of special creation.

The believer may still say, “God is behind the laws.” That is a different kind of claim. It is more abstract, more distant, and less testable. It no longer explains the specific event in the way older religion did.

This is precisely the point.

As science advances, God often moves from direct explanation to background belief.

That movement is the retreat.

Evidence-Based Response

The central issue is not whether anyone can still believe in God after science. Of course people can.

The issue is whether God remains necessary as an explanation.

In one field after another, the answer has become no.

The eclipse does not require divine warning.

The storm does not require divine anger.

The plague does not require divine punishment.

The king does not require divine appointment.

The human body does not require mystical anatomy.

The movement of planets does not require heavenly spirits.

The history of Earth does not require sacred chronology.

The diversity of life does not require separate creation of each species.

This does not prove that no God can exist in any philosophical sense. That larger argument will be developed across the series. But it does show that the old explanatory role of God has been steadily weakened.

The God who once explained everything now explains less and less.

That is not a small change. It is one of the greatest intellectual changes in human history.

Chapter Summary

Before science, God explained the world.

God explained the sky, weather, disease, disaster, birth, death, kingship, morality, and fate. These explanations were understandable because human beings were trying to survive in a frightening world without modern scientific tools.

Religion gave people comfort, meaning, ritual, authority, and a sense of control. But as an explanation of nature, it was often wrong. It mistook natural processes for divine actions. It turned disease into punishment, disaster into judgment, and celestial events into omens.

Science began to change this by asking a different question. Instead of asking what the gods wanted, science asked how nature worked.

That question opened the door to a new world.

Key Points

Before science, supernatural explanations felt natural because better explanations were not available.

Ancient people often interpreted the sky, disease, disaster, and political power through divine will.

Religion gave emotional comfort and social order, but it often failed as an explanation of nature.

Science became powerful because it looked for natural causes and allowed evidence to correct old beliefs.

The retreat of God began when human beings learned to ask not "Which god caused this?" but "What natural process explains this?"

Transition to the Next Chapter

If the old world placed God behind nearly everything, the next question is clear.

How did human beings begin to escape that way of thinking?

The answer begins not with a single discovery, but with a new habit of mind. Human beings had to learn to look carefully at nature, measure what they saw, test their ideas, and accept correction when they were wrong.

That habit became science.

In the next chapter, we will examine the first rule of science: look at nature.

Sources

Encyclopedia Britannica, "Omen"

Encyclopedia Britannica, "Divination"

Encyclopedia Britannica, "Germ Theory"

Stanford Encyclopedia of Philosophy, "Religion and Science"

World History Encyclopedia, "Ancient Greek Medicine"

World History Encyclopedia, "Roman Religion"

David Hume, *The Natural History of Religion*

James George Frazer, *The Golden Bough*

Émile Durkheim, *The Elementary Forms of Religious Life*

Mircea Eliade, *The Sacred and the Profane*

Pascal Boyer, *Religion Explained*

Daniel Dennett, *Breaking the Spell*

Carl Sagan, *The Demon-Haunted World*

Chapter 2: The First Rule of Science: Look at Nature

The Moment Human Beings Looked Again

For most of human history, people looked at nature through stories they had already inherited.

The sky was not simply sky. It was heaven, warning, judgment, blessing, or the home of gods. Disease was not simply disease. It was punishment, impurity, possession, curse, or fate. Disaster was not simply disaster. It was anger from above. A strange dream was not simply a dream. It might be a message from the dead or a sign from the divine.

Human beings did not lack intelligence. They lacked a reliable method.

They observed the world, but they often interpreted it before they truly examined it. They saw lightning and imagined anger. They saw fever and imagined sin. They saw eclipses and imagined cosmic danger. They saw unusual stars and imagined omens.

Then a new habit slowly appeared.

Look again.

Do not begin with the priest.

Do not begin with the sacred story.

Do not begin with the inherited answer.

Begin with nature.

What is actually happening?

That question sounds simple now, but it was revolutionary. It asked human beings to look before they concluded, measure before they believed, test before they trusted, and correct themselves when reality disagreed.

This was the beginning of science as a disciplined way of knowing.

Science did not become powerful because it had all the answers from the start. It became powerful because it had a better way of finding answers. It did not depend only on sacred books, priestly authority, royal command, ancient tradition, or emotional comfort. It asked nature itself.

What do we see?

What can we measure?

What happens again and again?

What prediction can we make?

What evidence would prove us wrong?

Can someone else repeat the observation?

Does this explanation work better than the older one?

These questions changed the human future.

They moved the mind from sacred certainty toward disciplined doubt. They moved explanation from heaven to earth. They made it possible to discover that many things once attributed to God, gods, demons, spirits, curses, or miracles had natural causes.

This was one of the greatest turning points in human history.

The Old Way: Begin With the Answer

The old religious way of explaining nature often began with the answer already known.

God did it.

The gods were angry.

A spirit caused it.

A demon entered the person.

A curse fell on the family.

A sacrifice was needed.

A sacred law was broken.

A divine plan was unfolding.

Once the answer was assumed, every event could be fitted into it.

If rain came after prayer, prayer worked.

If rain did not come, perhaps the prayer was not sincere enough.

If a sick person recovered, God healed him.

If the sick person died, God had another plan.

If a king won a war, God favored him.

If he lost, God punished him, tested him, or withdrew favor.

This made religious explanation emotionally flexible, but intellectually weak.

An explanation that can survive every outcome explains very little. If every possible result confirms the belief, then the belief is not really being tested. It is being protected.

Science slowly introduced a different discipline.

It said that an explanation must risk being wrong.

That is a difficult idea. Human beings do not like being wrong. We become attached to our beliefs, especially when they are tied to family, culture, morality, identity, fear, and hope. But reality does not become true because we love an idea. Nature does not change because a belief is ancient. Disease does not obey tradition. Planets do not follow scripture. Lightning does not respect authority.

The first rule of science is simple but powerful:

Look at nature.

Do not begin with the answer.

Begin with the evidence.

Seeing Is Not Enough

At first, this may sound easy.

Just look at the world.

But science is not merely looking. Human beings had been looking at the world since the beginning. They saw storms, eclipses, fever, birth, death, stars, plants, animals, rivers, mountains, and seasons. Yet they often misunderstood what they saw.

Why?

Because seeing is not the same as understanding.

A person can see lightning and believe it is a weapon of God. A person can see fever and believe it is punishment. A person can see the Sun move across the sky and believe the Sun travels around a fixed Earth. A person can see the body die and imagine a ghost leaving it.

The eyes give information, but the mind interprets it.

And the mind is not neutral.

The mind carries fear, expectation, culture, memory, religious teaching, authority, and desire. We often see what we have been trained to see. We interpret the world through the stories we already believe.

This is why science needed more than observation.

It needed careful observation.

It needed measurement.

It needed comparison.

It needed instruments.

It needed mathematics.

It needed experiment.

It needed repeated testing.

It needed the courage to say, "I was wrong."

Science is disciplined seeing.

It is not just looking at the world. It is learning how not to fool ourselves while looking.

This is crucial because human beings are very good at fooling themselves. We notice what supports our belief and ignore what challenges it. We remember the prayer that seemed to work and forget the many prayers that did not. We see patterns where none exist. We assign intention to events that have no intention. We prefer a comforting falsehood to a painful uncertainty.

Science does not remove all human bias, but it builds methods to fight against it.

That is why it became so successful.

Measurement: Turning Impressions Into Evidence

One of the great steps toward science was measurement.

Before measurement, people could say many things loosely.

The season is changing.

The fever is high.

The star is bright.

The river is rising.

The crop is poor.

The patient is worse.

The planet has moved.

These observations matter, but they are imprecise. Measurement makes them sharper.

How high is the fever?

How much has the river risen?

How far did the planet move?

How long did the eclipse last?

How many patients survived?

How many died?

How much medicine was given?

How often did the event repeat?

Measurement changes the quality of thought.

It moves us from impression to evidence. It makes comparison possible. It allows one person's claim to be checked by another person. It lets us find patterns too subtle for ordinary perception.

A thermometer is not merely a tool. It is a way of disciplining the mind. A clock, a scale, a telescope, a microscope, a ruler, a calendar, a laboratory instrument, and a statistical table all do something similar. They reduce the power of vague belief.

They ask reality to give a clearer answer.

This is why science often grows with instruments. The naked eye sees only so much. The telescope reveals moons and galaxies. The microscope reveals cells and microbes. The X-ray reveals bone and lung. The blood test reveals chemistry. The scanner reveals structure. The clock reveals timing. The scale reveals change.

Each instrument extends human perception.

Each extension of perception has reduced the territory of supernatural explanation.

When the microscope revealed microorganisms, disease began to move away from demons and curses. When the telescope revealed the structure of the heavens, the sky moved away from sacred perfection. When geological tools revealed layers of rock and fossils, Earth's history moved away from narrow sacred timelines.

Measurement is one of the quiet enemies of superstition.

Not because numbers are everything, but because numbers force us to answer to reality.

Prediction: The Test of Understanding

A powerful explanation does more than describe what already happened. It helps us predict what will happen.

This is one of the great strengths of science.

If we understand an eclipse, we can predict the next one.

If we understand the orbit of a planet, we can calculate where it will be.

If we understand infection, we can predict how disease may spread.

If we understand weather patterns, we can forecast storms.

If we understand anatomy and physiology, we can often predict what may happen when an artery is blocked, a lung collapses, or an organ fails.

Prediction is not always perfect. Science often deals with probability, complexity, and uncertainty. Weather forecasts can be wrong. Medical outcomes can vary. Biological systems are complex. But even imperfect prediction is very different from sacred interpretation.

A priest can say that an eclipse is a warning from heaven. But can he predict the exact time, path, and duration of the next eclipse?

A person can say disease is caused by sin. But can he predict who will become infected after exposure to a specific microbe?

A person can say a comet is an omen. But can he calculate its return?

The ability to predict is a sign that we are beginning to understand the mechanism.

This is where divine explanations often fail. "God did it" may sound like an answer, but it rarely predicts anything clearly. It does not tell us when the next earthquake will occur. It does not tell us which child will recover from infection. It does not tell us which volcano will erupt. It does not tell us how to prevent smallpox, cholera, tuberculosis, or sepsis.

Natural explanations are powerful because they connect cause and effect.

Once we know that contaminated water can spread cholera, we can predict danger and prevent disease. Once we know that mosquitoes transmit malaria, we can reduce transmission. Once we know that gravity governs motion, we can calculate trajectories. Once we know that the Moon passes between Earth and Sun, we can predict an eclipse.

Prediction turns explanation into power.

Not power over truth, but power to live better within reality.

Experiment: Asking Nature a Controlled Question

Observation watches nature.

Experiment questions nature.

This is another major step in scientific thinking. In ordinary life, many things happen at once. A patient receives medicine, rests, eats differently, is prayed for, and improves. What caused the improvement? Was it the medicine? Rest? Time? The natural course of the illness? The prayer? The change in diet? The body's immune response?

Without careful testing, we may choose the cause we prefer.

Experiment tries to separate causes.

It asks a focused question under controlled conditions. What happens if we change this one thing? What happens if we compare one group with another? What happens if we repeat the same test? What happens if the expected result does not occur?

This is why experiment was so dangerous to old authority.

A sacred claim may demand reverence. An experiment demands results.

If a treatment does not work, it should not survive merely because it is traditional. If a claim cannot be tested, we should not pretend it has the same strength as tested knowledge. If a belief produces no reliable result, we should not confuse emotional comfort with evidence.

This is especially important in medicine.

Patients often improve after many different actions. They may improve after prayer, herbs, rest, rituals, pills, injections, or no treatment at all. The human body heals many conditions naturally. Symptoms rise and fall. Pain fluctuates. Fever breaks. Cough improves. Wounds close.

Without careful comparison, people easily give credit to the wrong cause.

This is why modern medicine depends on evidence. A treatment must be tested. It must be compared. Benefits and harms must be studied. One dramatic story is not enough.

Religion often relies on testimony: “I prayed, and I recovered.”

Science asks a harder question: "How often do people recover with prayer compared with similar people without that prayer? What else was done? What was the disease? What would have happened naturally? Can the result be repeated?"

That may sound less emotional, but it is more honest.

Science does not reject human experience. It asks whether experience has been interpreted correctly.

When Science Works, Life Changes

Natural explanation proved itself by working.

This is one of the strongest arguments for science.

When astronomy understood eclipses, fear gave way to prediction.

When navigation used mathematics and the stars, ships crossed oceans more reliably.

When physics understood motion, bridges, engines, machines, and later spacecraft became possible.

When chemistry understood substances, medicines, fertilizers, antiseptics, and materials changed daily life.

When germ theory understood infection, sanitation, vaccination, sterilization, and antibiotics saved lives.

When anatomy and physiology understood the body, surgery became safer and diagnosis became more accurate.

These successes matter because they are not merely ideas on paper. They touch human life.

A mother whose child survives pneumonia because of antibiotics does not need a theory that merely sounds comforting. She needs an explanation that works.

A surgeon operating in a sterile room does not need ritual purity. He needs antisepsis, anatomy, anesthesia, and careful technique.

A pilot crossing the sky does not need an omen from the planets. He needs physics, weather science, instruments, and navigation.

A satellite in orbit does not remain there because of prayer. It remains there because engineers understand gravity, motion, velocity, mass, and calculation.

This is why science slowly became more powerful than supernatural explanation.

It delivered results.

It did not answer every question of the heart. It did not eliminate grief, death, loneliness, or moral struggle. But when it came to explaining nature, predicting events, and reducing suffering, it worked in ways religion did not.

That practical success changed the world.

Correction: The Moral Discipline of Science

The greatest strength of science is not that scientists are always right.

They are not.

Scientists can be wrong, stubborn, arrogant, biased, jealous, political, and limited by the culture of their time. Scientific institutions can resist new ideas. Old theories can survive too long. New theories can be rejected too quickly. Mistakes occur.

But science has something precious.

It has a method of correction.

A scientific claim must eventually face evidence. If evidence continues to contradict it, the claim weakens. If another explanation works better, the old explanation must yield. If an experiment cannot be repeated, confidence declines. If new instruments reveal new facts, the theory must adjust.

This is why science can improve even though scientists are human.

Science is not perfect. But it is self-correcting in a way that sacred certainty usually is not.

In religion, truth is often protected by authority. In science, truth is exposed to risk.

That difference matters deeply.

A religious doctrine may remain unchanged for centuries because it is considered holy. A scientific idea may be replaced when evidence demands it. A sacred text may be interpreted again and again to preserve belief. A scientific theory must answer to the world.

This does not mean science changes randomly. It changes because better evidence and better explanations appear. The process can be slow and messy, but over time, reality has a way of pressing against error.

This is why science has been able to build airplanes, vaccines, antibiotics, satellites, anesthesia, electricity, clean water systems, computers, telescopes, and modern surgery.

These things did not emerge from sacred certainty.

They emerged from tested knowledge.

Science works not because it never makes mistakes, but because it has a way to find and correct mistakes.

Natural Explanation

Science searches for natural explanations.

This does not mean science begins with hatred of God. It means science studies the world through causes that can be observed, tested, measured, compared, and corrected.

If lightning strikes, science looks for electrical causes.

If disease spreads, science looks for biological causes.

If a planet moves, science looks for physical causes.

If species change, science looks for natural mechanisms.

If memory changes after brain injury, science looks at the brain.

This natural approach is sometimes misunderstood. Some believers say, "Science excludes God unfairly." But science does not exclude God in the way a courtroom might exclude a witness. Science simply requires explanations that can do scientific work.

A scientific explanation must clarify a mechanism. It must connect cause and effect. It must allow testing or observation. It must produce reliable understanding. It must be open to correction.

"God did it" does not usually meet these requirements.

It does not tell us how lightning forms. It does not tell us which microbe causes tuberculosis. It does not tell us how blood circulates. It does not tell us why planets follow calculable paths. It does not tell us how inherited traits change in populations over time.

It may satisfy a religious mind, but it does not function as a scientific explanation.

This is why natural explanation became so successful. It did not win because it was emotionally richer. Often it was emotionally colder. It won because it could be tested, corrected, and used.

Natural explanation allowed human beings to act effectively.

If cholera is caused by divine punishment, repentance may seem the answer. If cholera is spread through contaminated water, clean water becomes the answer.

If mental illness is caused by possession, exorcism may seem the answer. If mental illness involves the brain, trauma, genes, chemicals, and environment, treatment becomes possible.

If famine is caused by angry gods, sacrifice may seem the answer. If famine involves rainfall, soil, pests, storage, economics, war, and distribution, practical action becomes possible.

Natural explanation does not merely explain.

It liberates.

It frees human beings from false causes.

The Difference Between Meaning and Mechanism

One reason people resist scientific explanation is that they fear it removes meaning.

If disease is caused by microbes, does suffering have no meaning?

If the Earth is not the center of creation, are human beings worthless?

If life evolved, is life less precious?

If the mind depends on the brain, is love only chemistry?

These fears are understandable, but they confuse two different kinds of questions.

Science explains mechanism.

It asks how something happens.

Religion often gives meaning.

It asks why this matters, what purpose it serves, what moral lesson it contains, or how we should emotionally understand it.

The problem begins when religion mistakes meaning for mechanism.

A grieving person may find meaning in prayer. That is a human experience. But prayer does not explain the biological cause of pneumonia. A community may find meaning in ritual after a disaster. But ritual does not explain tectonic plates, weather systems, or infection. A person may see moral lessons in suffering. But those lessons do not identify the physical cause of suffering.

Meaning and mechanism are different.

This book is mainly about mechanism.

It is about how the natural world works and how science replaced divine explanations of that world. It does not claim that scientific mechanism answers every emotional or moral question. It does claim that when we are explaining nature, mechanism matters.

If we confuse meaning with mechanism, we may comfort ourselves but fail to understand reality.

And when we fail to understand reality, people suffer.

The Courage to Say "I Do Not Know"

Science begins with humility.

Not fake humility. Real humility.

It says, "I do not know, so I must investigate."

This is very different from pretending to know.

For much of history, religious explanations gave quick answers. They offered certainty where there was mystery. They gave names to invisible causes. They filled silence with divine intention.

Science often begins with a less satisfying sentence:

"I do not know."

But that sentence is powerful.

"I do not know" leaves room for discovery. "God did it" often ends the search too early. "I do not know" invites observation, measurement, experiment, and correction. "God did it" may turn mystery into doctrine.

There is dignity in not knowing.

There is honesty in admitting uncertainty.

There is wisdom in refusing to fill every gap with imagination.

This is one of the great moral lessons of science. It teaches intellectual patience. It teaches us not to claim more than we know. It teaches us to be careful with certainty.

That is especially important when certainty has consequences.

If we are certain that disease is punishment, we may blame the sick. If we are certain that disaster is divine judgment, we may condemn the victims. If we are certain that mental illness is possession, we may harm the suffering person. If we are certain that our scripture contains the final answer to nature, we may resist discovery.

False certainty can be cruel.

Scientific humility can be compassionate.

Why Science Threatened Sacred Authority

Science threatened religion not because every scientist hated religion, but because science created an independent path to truth.

That was the real danger.

If truth can be discovered by observing nature, then priests do not control all truth.

If the motion of planets can be calculated, then sacred cosmology can be wrong.

If the body can be understood through anatomy, then mystical ideas of the body can be corrected.

If disease can be explained by microbes, then moral and demonic explanations lose authority.

If Earth's history can be read in rocks, then sacred timelines can be challenged.

Science did not merely add facts. It changed who had authority to speak about reality.

The authority moved from tradition to evidence.

This shift was not easy. It produced conflict, fear, resistance, reinterpretation, and sometimes punishment. But once evidence became a source of truth independent of religious authority, the world could not return fully to the older way.

The telescope did not ask permission from theology.

The microscope did not ask permission from scripture.

The fossil did not ask permission from sacred chronology.

The experiment did not ask permission from tradition.

Nature became a book that could be read directly.

And once human beings began reading that book carefully, many older religious explanations began to fade.

Science as a Human Achievement

Science is sometimes described as cold, mechanical, or anti-human. That is a mistake.

Science is one of the most human achievements in history.

It comes from curiosity, patience, doubt, imagination, discipline, and courage. It requires the humility to admit ignorance and the confidence to search anyway. It requires cooperation across generations. It requires trust, criticism, honesty, and correction.

No single person created science. It grew slowly across cultures and centuries. Ancient observers studied the heavens. Physicians recorded symptoms. Mathematicians developed tools. Philosophers asked about nature. Craftsmen built instruments. Navigators needed accurate maps and skies. Anatomists studied bodies. Chemists studied matter. Natural philosophers searched for order.

Science was not born perfect. It emerged gradually.

But over time, it developed a habit that no older system of explanation had used with the same power:

Let reality correct belief.

That habit made science different.

It allowed knowledge to grow beyond tribe, scripture, and tradition. It allowed one generation to improve on another. It allowed human beings to admit error without destroying the entire structure of knowledge.

A scientific correction is not a disaster.

It is progress.

This is why science could move from astrology to astronomy, from alchemy to chemistry, from demons to disease mechanisms, from sacred geography to geology, from special creation to evolution, and from soul-language to brain science.

Science did not arrive all at once. But once its method matured, the old divine explanations could not remain untouched.

The Religious Reply

A thoughtful believer may say:

"Science explains the natural process, but God created the process. Science explains how things happen, but God explains why there is anything at all. Natural laws do not remove God. They may simply describe the way God governs the universe."

This is a serious reply. It should not be dismissed carelessly.

Many religious thinkers have taken this position. They do not deny science. They accept astronomy, physics, medicine, and biology, but they place God behind the natural order. For them, God is not the immediate cause of each storm, eclipse, infection, or planetary motion. God is the deeper source of the whole system.

This view is more sophisticated than the older idea that gods directly throw lightning, send plagues, move planets, or punish villages with drought.

But it also shows how far the argument has shifted.

God has moved from direct explanation to background interpretation.

That is a major retreat.

Once we understand lightning through atmospheric electricity, infection through microbes, and planetary motion through gravity, God is no longer doing the specific explanatory work. God has become a philosophical or theological addition placed behind the natural explanation.

A person may still believe that.

But science does not need it.

Evidence-Based Response

The issue is not whether someone can add God behind natural law. The issue is whether God is necessary to explain what we observe.

In scientific terms, an explanation should do work. It should clarify the cause. It should improve prediction. It should guide investigation. It should distinguish one possible outcome from another. It should help us understand reality better.

Saying “God is behind the laws” may be personally meaningful, but it usually does not improve the scientific explanation.

It does not help calculate the eclipse.

It does not identify the pathogen.

It does not predict the orbit.

It does not explain the fossil layer.

It does not improve the treatment of infection.

It does not show how blood circulates.

It does not tell us why one theory is better than another.

Science can proceed without that assumption.

That is what “unnecessary” means in this book.

It does not mean every religious belief disappears. It does not mean no person can believe in God. It means that God is no longer required for the explanation itself.

This is why the first rule of science changed history.

Look at nature.

Let evidence speak.

Test the claim.

Correct the error.

Do not protect an answer simply because it is old, comforting, or sacred.

That method steadily reduced the need for divine explanation.

Chapter Summary

Science began with a new habit of mind.

Instead of beginning with sacred answers, it began with observation. Instead of relying only on authority, it relied on evidence. Instead of protecting every inherited belief, it allowed correction. Instead of asking what the gods wanted, it asked how nature worked.

This method changed everything.

Observation, measurement, prediction, experiment, and correction allowed human beings to explain eclipses, disease, planetary motion, anatomy, chemistry, geology, and life without invoking divine action.

Science did not become powerful because scientists were always right. It became powerful because it had a method for discovering when it was wrong.

That self-correcting method made God less and less necessary as an explanation of nature.

Key Points

Science began with the disciplined habit of looking carefully at nature.

Observation alone is not enough, because human beings often misinterpret what they see.

Measurement turns vague impressions into evidence.

Prediction tests whether an explanation truly understands the mechanism.

Experiment asks nature controlled questions.

Science is powerful because it can correct itself.

Natural explanation does not deny meaning, but it separates mechanism from emotional interpretation.

God becomes unnecessary when natural causes explain the evidence.

Transition to the Next Chapter

Once human beings learned to look at nature more carefully, the first great battlefield was the sky.

For centuries, Earth seemed fixed, central, and special. The heavens seemed to revolve around us. This view was emotionally satisfying and religiously powerful. It placed humanity near the center of creation.

Then came a quiet but dangerous idea.

What if Earth was not the center?

In the next chapter, we will meet Copernicus and the beginning of one of the greatest displacements in human history.

Sources

Aristotle, *Physics*

Francis Bacon, *Novum Organum*

Galileo Galilei, *Dialogue Concerning the Two Chief World Systems*

Isaac Newton, *Mathematical Principles of Natural Philosophy*

Thomas Kuhn, *The Structure of Scientific Revolutions*

Karl Popper, *The Logic of Scientific Discovery*

John Losee, *A Historical Introduction to the Philosophy of Science*

Alan Chalmers, *What Is This Thing Called Science?*

John Hedley Brooke, *Science and Religion: Some Historical Perspectives*

Peter Harrison, *The Territories of Science and Religion*

David Wootton, *The Invention of Science*

Carl Sagan, *The Demon-Haunted World*

Jacob Bronowski, *The Ascent of Man*

Chapter 3: When Earth Lost the Center

The Comfort of a Still Earth

For most of human history, the Earth did not feel like a planet.

It felt like the ground of reality.

People stood on it, walked on it, built homes on it, grew food from it, buried their dead in it, and prayed beneath the sky above it. The Earth seemed solid, still, dependable, and central. It was the stage on which human life unfolded.

Above it was the sky.

The Sun rose in the east and set in the west. The Moon traveled across the night. The stars turned overhead in great circles. The planets wandered slowly among them. Everything seemed to move around us.

The evidence of ordinary experience looked obvious.

The Earth was fixed.

The heavens moved.

Humanity stood near the center.

This was not only an astronomical idea. It was emotionally satisfying. A still and central Earth gave human beings a meaningful place in the universe. The world seemed arranged around life, prayer, morality, kingship, harvest, judgment, and salvation.

A central Earth made religious imagination feel natural.

If human beings were central in God's plan, why should the Earth not be central in the cosmos? If heaven was above and the underworld below, then the universe itself seemed morally arranged. If the stars and planets moved over us, perhaps they carried messages, warnings, or signs.

The old sky was not empty space.

It was sacred architecture.

To move the Earth was therefore not a small correction. It was not merely a technical adjustment in astronomy. It disturbed an entire way of feeling at home in the universe.

When Copernicus placed the Sun near the center and made Earth a moving planet, he did far more than change a diagram.

He shook the human throne.

A World Built Around Us

The ancient Earth-centered universe did not begin as a foolish idea.

It was based on what people seemed to observe. The ground did not appear to move. A stone dropped from the hand fell straight down. Birds did not seem left behind by a spinning Earth. The air did not rush past like a violent wind. The Sun, Moon, stars, and planets appeared to move across the sky.

To the unaided eye, Earth seemed stationary.

The great Greek astronomer Ptolemy gave the ancient world a powerful mathematical model of this Earth-centered cosmos. In his system, Earth stood at the center while the Moon, Sun, planets, and stars moved around it in complex paths. The model was not simple, but it was useful. It could describe and predict planetary positions with reasonable success for its time.

This is important.

The old model did not survive merely because priests liked it. It survived because it seemed to work well enough. It fit ordinary experience. It fit common sense. It fit inherited philosophy. It also fit religious imagination.

In the Christianized medieval world, this Earth-centered universe became woven into theology, philosophy, education, and culture. Not every detail came directly from scripture, but the general picture felt comfortable to many religious minds. Earth was the stage of human drama. Heaven was above. Hell was below. The cosmic order seemed linked to moral order.

This made the old system powerful.

It was not just a scientific model. It was a complete worldview.

And worldviews do not fall easily.

A person can change an opinion quickly. But a worldview is different. A worldview tells people where they are, what matters, who has authority, and how the visible world connects to the invisible one. When a worldview is challenged, people often feel that reality itself is being threatened.

That is why Copernicus mattered.

He did not simply suggest that a planet moved.

He suggested that the human world was not where people thought it was.

Copernicus and the Dangerous Simplicity

Nicolaus Copernicus was not a loud revolutionary in the modern sense.

He was a careful scholar, a church canon, a mathematician, and an astronomer. His great work, *On the Revolutions of the Heavenly Spheres*, was published in 1543, the year of his death.

The idea at its heart was simple but dangerous.

Earth moves.

Earth rotates daily on its axis. Earth travels yearly around the Sun. The apparent motion of the heavens is partly caused by our own motion.

This was a profound reversal.

For centuries, people had looked up and assumed that the heavens moved around them. Copernicus asked them to consider something deeply unsettling: perhaps the motion we see in the sky is partly the result of the motion we do not feel beneath our feet.

This is one of the great lessons of science.

Reality is not always what it feels like.

The Earth feels still, but it moves.

The Sun appears to travel around us, but that appearance is misleading.

The sky seems arranged around us, but we are not the center.

Science often begins by correcting the arrogance of ordinary perception.

Copernicus did not immediately prove everything in the modern sense. His model still used circular motion. It was not yet Newtonian physics. It did not instantly persuade everyone. But it changed the question forever.

Before Copernicus, the natural assumption was that Earth stood fixed at the center.

After Copernicus, that assumption could no longer be taken for granted.

A door had opened.

Once opened, it could not be fully closed.

The Emotional Shock of a Moving Earth

Today, children learn that Earth goes around the Sun. Diagrams show the solar system. Globes sit in classrooms. Photographs from space show Earth as a blue planet. We are used to the idea.

But we should not forget how strange it once sounded.

Imagine being told that the ground beneath your feet is moving at enormous speed through space. Imagine being told that the rising and setting of the Sun is not what it appears to be. Imagine being told that Earth is not the center, but one planet among others.

This was not only difficult intellectually. It was unsettling emotionally.

Human beings do not easily surrender centrality.

We like to feel important. We like to feel chosen. We like to believe that the world is arranged around our concerns. Religion often strengthens this feeling by presenting the universe as a stage for human salvation, divine judgment, moral testing, and eternal destiny.

The Copernican revolution weakened that comfort.

It did not prove atheism. It did not disprove God. Copernicus himself was not trying to create a godless universe. But the emotional direction of his discovery was clear.

Humanity became less central.

This is one of the great themes of modern knowledge.

Science has repeatedly shown that human beings are not as central as they once imagined.

Earth is not the center of the cosmos.

The Sun is not the center of the universe.

Our galaxy is one among many.

Human beings are one branch of life, not a separate creation outside nature.

The mind itself depends on the brain, not on an obvious immaterial soul.

Copernicus began this great displacement.

He did not finish it. But he started the movement.

Sacred Order and Cosmic Geography

Older religious imagination often depended on cosmic geography.

Heaven was above. Earth was here. Hell or the underworld was below. The physical universe was imagined in layers, and those layers often carried moral meaning. Higher meant purer, nobler, and closer to the divine. Lower meant heavier, darker, and more corrupt.

This way of thinking felt natural because it matched human experience. Light came from above. Rain came from above. Birds flew above. Smoke rose upward. The dead were buried below. Darkness was associated with caves, graves, and the earth beneath.

The universe was not only a place.

It was a moral diagram.

The Earth-centered cosmos supported that diagram. It placed human life inside a meaningful structure. Above us were the heavens. Around us were the celestial spheres. Beyond them was divine reality.

A moving Earth disturbed this symbolic order.

If Earth is a planet, then "above" and "below" are not absolute in the old sense. If Earth turns, then heaven is not simply a fixed roof over human life. If Earth orbits the Sun, then the cosmic architecture of everyday religious imagination begins to loosen.

Again, this does not disprove God.

But it changes the kind of universe in which God is imagined.

The older God lived comfortably in a small, layered, human-centered cosmos. The emerging scientific universe was larger, less personal, less obviously moral, and less centered on us.

That was the deeper challenge.

Religion could adapt, and often did. Theologians could reinterpret. Believers could say that God created a moving Earth just as easily as a fixed Earth. They could say scripture was not meant to be an astronomy textbook. They could preserve belief by making it more flexible.

But each adaptation carried a cost.

The more religion adjusted to science, the more it admitted that science had become the better guide to nature.

Why the Old Model Was Hard to Abandon

The Copernican model did not immediately sweep away the old Earth-centered view.

There were several reasons.

First, the old model was supported by common sense. People did not feel Earth moving. Without modern physics, it was not obvious how Earth could rotate and orbit without producing chaos.

Second, the old model had intellectual authority. Aristotle and Ptolemy were not ordinary names. They had shaped centuries of thought. Their ideas were embedded in education, philosophy, and theology.

Third, the old model had religious comfort. It seemed to fit a universe built around human beings and divine purpose.

Fourth, Copernicus's own model was not perfect. It still required technical adjustments. It did not yet have Kepler's ellipses or Newton's gravity. It was a major step, but not the final system.

This is an important lesson.

Scientific revolutions are often gradual.

A new idea may be closer to truth, but still incomplete. It may explain some things better while leaving other problems unresolved. It may require later thinkers to refine, correct, and strengthen it.

Copernicus began the change. Galileo, Kepler, Newton, and many others would carry it forward.

This is how science often works.

One person does not suddenly possess final truth. Instead, one discovery weakens an older belief. Another discovery strengthens the new direction. A later theory explains why the new model works. Evidence accumulates. Resistance slowly becomes harder to maintain.

God does not retreat all at once.

The retreat happens through stages.

From Center to Planet

The Copernican revolution began the end of cosmic privilege.

Cosmic privilege means the feeling that human beings occupy a special central place in the physical universe. The old Earth-centered system supported that feeling. It allowed people to imagine that the universe was arranged around human life.

Copernicus made that harder.

If Earth is a planet, then it is not the fixed center. If Earth moves, then the heavens do not simply revolve around us. If our position is not central, then perhaps our importance cannot be based on location.

This was a serious blow to religious imagination, even if not always admitted.

Religion often gives human beings a privileged place. We are made in the image of God. We are the center of divine concern. The universe is created with us in mind. Our sins shake heaven. Our prayers reach the creator of all things. Our destiny lasts forever.

The moving Earth did not directly refute these claims. But it made them feel less obvious.

A small planet moving around a star does not look like the fixed platform of cosmic importance.

Later astronomy would deepen this displacement. The Sun itself would become one star among many. The Milky Way would become one galaxy among billions. The visible universe would become unimaginably vast.

But the first wound to centrality came with Copernicus.

After him, humanity could no longer honestly look at the sky in the same way.

The Sky Without Human Flattery

One of the quiet gifts of science is that it removes human flattery from nature.

This can feel painful at first.

People often want the universe to confirm their importance. They want the sky to speak to them. They want events to contain personal messages. They want the cosmos to be morally arranged around human hopes and fears.

Science offers something different.

It says the universe is not arranged to flatter us.

The planets do not move for our spiritual drama. The stars do not exist mainly as signs for human affairs. The Sun does not rise because we need symbolism. The Moon does not change shape to deliver messages about our fate.

Nature is not a mirror of human importance.

This can sound cold, but it is also liberating.

If the sky is not sending omens, we do not need to live in fear of every comet, eclipse, or unusual light. If planets are not controlling our destiny, we can take responsibility for our lives. If Earth is not the center, we can learn humility. If the universe is vast and indifferent, human kindness becomes more important, not less.

Science removes false comfort, but it also removes false fear.

The Copernican revolution helped begin that process.

It taught humanity that the sky is not a sacred ceiling built around us. It is a natural system that can be studied, measured, predicted, and understood.

That was a great loss for superstition.

It was a great gain for truth.

Natural Explanation Proves Itself

Copernicus did not merely offer a new religious interpretation.

He offered a new model of the heavens.

That matters.

A scientific model is not strong because it is emotionally satisfying. It becomes strong when it explains more, organizes evidence better, raises better questions, and opens the way to further discovery.

The Copernican model did this.

It made it possible to think about the apparent motion of the heavens in a new way. It helped later astronomers understand planetary motion more deeply. It prepared the way for Galileo's observations, Kepler's laws, and Newton's gravitational universe.

Science often grows this way.

One idea opens a path.

The first version may be incomplete. It may still carry old assumptions. It may need correction. But it changes the direction of inquiry.

This is very different from sacred certainty.

A sacred cosmology may defend the old picture because it feels meaningful. A scientific model must eventually answer to evidence. Copernicus did not end the debate, but he shifted the burden. From that point forward, the Earth-centered universe had to defend itself against a new possibility.

The Earth might move.

Once that thought entered serious astronomy, the old religiously comfortable universe began to lose its authority.

The Religious Reply

A thoughtful believer may respond:

"Copernicus did not disprove God. Earth can move around the Sun and still be created by God. A human being does not need to live at the physical center of the universe to be spiritually important. God's concern for humanity does not depend on astronomy."

This is a strong reply, and it is partly correct.

Copernicus did not disprove God. A moving Earth is not a direct argument against every form of belief. A religious person can believe that God created a heliocentric solar system. Human moral worth does not logically depend on being physically central.

But this reply also shows how religious belief adapts after science forces a change.

The older worldview felt comfortable because cosmic structure and religious meaning seemed to support each other. Earth's centrality fit the feeling of human importance. Heaven above and Earth below fit sacred imagination. The movement of the heavens seemed to circle the human world.

Once science changed the astronomy, religion had to reinterpret the meaning.

This reinterpretation may preserve belief, but it gives up explanatory authority over the sky. Religion no longer tells us how the heavens are arranged. Science does. Religion must now say that the arrangement does not matter spiritually.

That may be emotionally possible.

But it is still a retreat from the older view.

Evidence-Based Response

The Copernican revolution did not prove that God does not exist.

But it weakened one of the oldest and most comforting assumptions of religion: that the physical universe is centered around human beings.

It also showed something even more important.

Human experience can be deeply misleading.

The Earth feels still, but it moves.

The Sun appears to travel around us, but the appearance is deceptive.

The heavens seem to revolve around human life, but they do not.

This was a lesson in humility.

It taught human beings that reality is not obligated to match our intuition, our tradition, or our religious comfort. If evidence shows that Earth moves, then Earth moves, no matter what common sense, ancient authority, or sacred imagination may suggest.

That lesson would echo through the rest of science.

The body would not be what ancient medicine imagined. Disease would not be what priests and healers often claimed. Earth's age would not fit narrow sacred chronologies. Life would not be explained by separate acts of creation. The universe would not remain small, young, and human-centered.

Copernicus opened the door to a larger truth:

Nature must be understood on its own terms.

Not on the terms of human comfort.

Not on the terms of inherited authority.

Not on the terms of religious expectation.

On its own terms.

Chapter Summary

For most of human history, Earth seemed fixed, central, and special. The Sun, Moon, planets, and stars appeared to move around it. This Earth-centered view fit ordinary experience, ancient philosophy, and religious imagination.

Copernicus challenged that picture by proposing that Earth rotates and orbits the Sun. His model did not immediately solve every astronomical problem, but it changed the direction of thought forever.

The Copernican revolution was not only about astronomy. It was about human centrality. It weakened the comforting belief that the universe was physically arranged around us. It showed that ordinary perception can be misleading and that inherited authority can be wrong.

Copernicus did not disprove God. But he began the long process of removing humanity from the center of the cosmos.

That was one of the first great retreats of divine explanation.

Key Points

The Earth-centered universe felt natural because it matched ordinary human experience.

The old cosmic model also supported religious imagination by placing humanity near the center of creation.

Copernicus proposed that Earth moves, rotating on its axis and orbiting the Sun.

This did not disprove God, but it weakened the idea of a human-centered cosmos.

The Copernican revolution showed that reality may differ deeply from common sense and religious comfort.

This was the beginning of humanity's cosmic displacement.

Transition to the Next Chapter

Copernicus moved the Earth in theory.

But a theory, especially one that challenges sacred tradition, needs evidence that people can see.

That evidence would come through a new instrument pointed toward the sky: the telescope.

In the next chapter, we will meet Galileo and see how the heavens themselves lost their imagined perfection.

Sources

Nicolaus Copernicus, *On the Revolutions of the Heavenly Spheres*

Thomas S. Kuhn, *The Copernican Revolution*

Owen Gingerich, *The Book Nobody Read*

Encyclopedia Britannica, "Nicolaus Copernicus"

Stanford Encyclopedia of Philosophy, "Copernicus"

Stanford Encyclopedia of Philosophy, "Religion and Science"

Arthur Koestler, *The Sleepwalkers*

Jacob Bronowski, *The Ascent of Man*

Carl Sagan, *Cosmos*

John Hedley Brooke, *Science and Religion: Some Historical Perspectives*

Chapter 4: Galileo and the Fall of the Perfect Heavens

The Night the Heavens Changed

There are moments in history when a simple instrument changes the human mind.

The telescope was one of them.

Before the telescope, the sky belonged partly to the eye and partly to imagination. Human beings could see the Sun, the Moon, the stars, and the wandering planets. But they could not see enough to correct many ancient beliefs. The heavens looked smooth, distant, perfect, and unreachable. They seemed different from the rough and broken world below.

Earth was the place of dust, decay, birth, death, disease, conflict, and change.

The heavens were imagined as higher, purer, more permanent, and closer to the divine.

Then Galileo lifted a tube of glass toward the sky.

That simple act changed everything.

Through his telescope, the Moon was no longer a smooth heavenly lamp. It had mountains, valleys, shadows, and rough surfaces. Jupiter was no longer just a wandering light. It had moons of its own. The Sun was not a flawless divine disk. It had spots. Venus showed phases, changing shape in a way that made far better sense if it moved around the Sun.

The heavens were not perfect.

They were physical.

They had surfaces, shadows, motions, blemishes, and bodies moving around other bodies.

That was the shock.

Galileo did not merely see new objects. He damaged an old sacred picture of the universe. He showed that the sky was not a separate divine realm of perfect purity. It was part of nature.

The heavens had entered the world of evidence.

The Old Dream of Perfect Heavens

For many centuries, educated people in Europe inherited a view of the universe shaped by Aristotle, Ptolemy, and later Christian theology.

In that older picture, the universe was divided into two broad regions.

The earthly region was the lower world. It was changeable, heavy, corruptible, and imperfect. Here, things were born, aged, decayed, and died. Bodies fell.

Wood burned. Flesh became sick. Metal rusted. Plants grew and withered. Human beings suffered.

The heavenly region was different. It was thought to be more perfect, more orderly, more circular, more permanent, and less subject to decay. The celestial bodies moved in elegant patterns. The Moon, Sun, planets, and stars seemed to belong to a higher order of reality.

This distinction was not merely scientific. It carried emotional and religious force.

The heavens felt closer to God. They seemed pure because they were far above the messiness of earthly life. They seemed eternal because they changed so little to the naked eye. They seemed divine because they were luminous, orderly, and unreachable.

Human beings often confuse distance with perfection.

The faraway mountain looks smooth until we climb it. The distant hero looks flawless until we know him closely. The heavens looked perfect because no one had seen them closely enough.

The telescope ended that illusion.

Galileo's Dangerous Tube of Glass

Galileo Galilei did not invent the first telescope, but he improved the instrument and used it with extraordinary intelligence. In 1609, he began observing the heavens with telescopes that magnified far beyond ordinary vision. In 1610, he published his discoveries in *Sidereus Nuncius*, often translated as *The Starry Messenger*.

The title itself was fitting.

The stars had sent a message, but not the message ancient religion expected.

The message was this: the heavens are not what we imagined.

Galileo's telescope revealed a Moon with roughness and relief. He saw bright and dark areas changing with the angle of sunlight. He interpreted these as mountains, valleys, and shadows. The Moon looked less like a perfect heavenly sphere and more like another world.

This was a powerful blow to the old division between Earth and heaven. If the Moon had mountains and valleys, then it was not made of some flawless celestial substance. It was not a perfect divine ornament. It was a physical body.

Then came Jupiter.

In January 1610, Galileo observed small points of light near Jupiter. Over several nights, he noticed that these points changed position. They did not behave like fixed stars. He concluded that they were moons moving around Jupiter. Today, these are known as the Galilean moons: Io, Europa, Ganymede, and Callisto.

This mattered deeply.

If Jupiter had moons, then not everything revolved around Earth.

Earth was no longer the obvious center of all motion. There could be smaller systems within the larger system. A planet could carry moons while also moving through space.

The old picture was cracking.

The Moon Became a World

Galileo's Moon may be the most emotionally powerful part of his discovery.

For generations, people had looked at the Moon as a bright, silent object in the sky. It marked time. It lit the night. It guided travelers. It inspired calendars, myths, songs, rituals, and poetry. It seemed heavenly because it was distant and luminous.

Through the telescope, that heavenly object became strangely familiar.

It had shadows.

It had uneven surfaces.

It had depth.

It looked like land.

This did not make the Moon less beautiful. It made it more real. The Moon was no longer merely a symbol hanging above human life. It was a place, a body, a world with its own physical features.

That change matters.

When an object becomes physical, it becomes open to explanation. It can be observed, mapped, compared, measured, and studied. It no longer belongs mainly to sacred imagination. It belongs to nature.

This is one of the great movements in this book.

The sky becomes astronomy.

The body becomes anatomy.

Matter becomes chemistry.

Disease becomes microbiology.

Earth becomes geology.

Life becomes evolution.

The mind becomes neuroscience.

Galileo's Moon was one of the first dramatic examples.

A sacred-looking object became a natural object.

The Sun Was Not Perfect

The Sun had long been a symbol of divine brilliance, purity, life, and power.

Yet telescopic observations showed sunspots.

The Sun was not an unblemished disk. It had dark marks. These marks changed. They appeared, moved, and disappeared. The Sun itself seemed to rotate.

This was not what the older idea of perfect heavens expected.

A spotted Sun was a disturbing thought.

The religious meaning of the Sun varied across cultures, but almost everywhere it carried deep symbolic force. It gave light. It gave warmth. It marked time. It sustained crops. It was associated with life, vision, order, and majesty. To see marks on the Sun was to see imperfection in the very object that seemed most glorious.

Of course, sunspots did not disprove God.

But they did weaken a particular kind of religious and philosophical imagination: the belief that the heavens were perfect and fundamentally different from Earth.

The telescope made the heavens more earthly.

That was Galileo's great offense against the old view.

He did not merely argue.

He showed.

And showing can be more dangerous than arguing.

Venus and the Hidden Architecture of the Sky

Galileo's observations of Venus were also important.

Venus shows phases, somewhat like the Moon. It appears as a crescent, half-lit, or nearly full depending on its position relative to Earth and the Sun. These phases were difficult to reconcile with the older Ptolemaic system in its traditional form. They made much better sense in a system where Venus orbited the Sun.

This gave strong support to the Copernican direction of thought.

The sky was beginning to reveal its hidden architecture.

To modern readers, this may seem like a technical matter. But at the time, it was a serious challenge to inherited authority. The Earth-centered universe was not merely an astronomical model. It was tied to ancient philosophy, education, theology, and common sense.

Galileo's observations forced people to choose between what had long been taught and what could now be seen.

That is one of the turning points in human history.

When evidence becomes visible, authority becomes vulnerable.

The telescope did not care who held power. It did not care what Aristotle had written. It did not care what theologians preferred. It did not care what seemed emotionally comfortable.

It magnified light.

And that light carried evidence.

The Conflict Was About Authority

The Galileo story is often told too simply.

Some tell it as pure science against pure religion. Others soften it so much that the conflict almost disappears.

The truth is more complex.

Galileo lived in a religious world. Many intellectuals of his time were religious. The Church was not opposed to every form of astronomy. The relationship between science and religion was not one long simple war.

But the Galileo affair still matters because it revealed a profound tension.

Who has the authority to describe nature?

If scripture, tradition, and Church interpretation say one thing, but observation through a telescope suggests another, which should yield?

Galileo argued, in effect, that nature must be read through evidence. The book of nature and the book of scripture could not truly contradict each other, but when they seemed to do so, perhaps scripture had been interpreted wrongly in matters of nature.

This was dangerous.

It meant religious authority could be mistaken when speaking about the physical world.

That idea threatened more than one astronomical claim. It threatened a whole structure of authority. If the Church could be wrong about the arrangement of the heavens, where else might inherited authority be wrong? If observation could correct theology in astronomy, could evidence also correct inherited beliefs about disease, Earth history, life, and the human mind?

Galileo's telescope raised these questions, even when people were not ready to face them.

The Trial and the Larger Lesson

Galileo eventually faced the Roman Inquisition. In 1633, he was tried and forced to recant his support for the motion of Earth. He spent the rest of his life under house arrest.

This event became one of the most famous symbols in the history of science.

We should not treat it as the whole story of science and religion. That would be too crude. But we also should not dismiss its meaning. A man had looked through a telescope, gathered evidence, argued for a new understanding of the heavens, and was punished because his conclusions challenged religious authority.

The larger lesson is clear.

Sacred authority often has difficulty accepting correction from evidence.

This does not mean every religious person rejects evidence. Many do not. It does not mean every scientific person is brave or honest. Many are not. Human beings are complicated.

But the method matters.

Science says that claims about nature must answer to nature.

Religious authority often says that claims must answer to doctrine, tradition, or scripture.

When those two standards conflict, intellectual honesty requires that the natural claim be judged by evidence, not by theological comfort.

If the Moon has mountains, it has mountains.

If Jupiter has moons, it has moons.

If Venus has phases, it has phases.

If the Sun has spots, it has spots.

No authority can make evidence disappear.

Why Galileo Was So Threatening

Galileo was threatening because he made the invisible visible.

A philosophical argument can be debated endlessly. A mathematical model can be dismissed as a useful fiction. But a telescope gives the eye something to confront.

Look.

That was Galileo's power.

Look at the Moon.

Look at Jupiter.

Look at Venus.

Look at the Sun.

Look at the stars.

The telescope did not answer every question, but it expanded the court of evidence. It allowed human beings to see beyond ordinary sight. It made the universe less dependent on ancient imagination.

That is why instruments are so important in the history of science.

The telescope weakened sacred astronomy. The microscope would later weaken supernatural explanations of disease. The thermometer would discipline vague impressions of fever. The stethoscope would allow physicians to hear internal signs. The X-ray would reveal hidden structure. The laboratory would make invisible chemistry measurable.

Instruments do not merely collect information.

They discipline belief.

They show us that ordinary perception is limited and that inherited explanation may be wrong.

Galileo's telescope was one of the first great instruments in the retreat of divine explanation. It did not remove God from all minds. It removed a certain kind of heaven from serious astronomy.

The perfect heavens were gone.

The Heavens Become Physical

Once the telescope revealed the heavens as physical, something changed forever.

The Moon became a world.

Jupiter became a system.

Venus became evidence.

The Sun became an object with change and blemish.

The stars became more numerous than the naked eye had ever imagined.

The sky was no longer mainly a sacred ceiling above human life. It was a vast natural realm filled with bodies that could be studied.

This is one of the great movements in the history of thought: the transformation of sacred space into physical space.

The old religious imagination often divided the universe into the ordinary and the divine, the low and the high, the corrupt and the perfect. Galileo helped weaken that division. The same kind of careful observation that could be applied to earthly objects could also be applied to heavenly ones.

The heavens were no longer exempt from investigation.

That was revolutionary.

Once the heavens became physical, they became part of science. Once they became part of science, divine explanation lost ground. A planet was no longer a wandering divine sign. A moon was no longer a perfect heavenly ornament. A star was no longer simply a symbol in a sacred drama.

These things became objects in nature.

And nature could be studied without invoking God.

Natural Explanation Proves Itself

Galileo's observations did more than disturb old beliefs.

They opened the way to a more reliable understanding of the sky.

The telescope showed that ordinary sight was not enough. The unaided eye had misled human beings for centuries. The heavens looked smooth and perfect because human vision was limited. Once the instrument extended the eye, the old explanation weakened.

This is a key lesson.

Natural explanation often begins when a better tool reveals a deeper reality.

The telescope revealed celestial bodies. The microscope would later reveal microbes. The laboratory balance would reveal chemical conservation. The fossil record would reveal deep time. Brain imaging would reveal patterns of structure and function.

Each tool made reality less dependent on inherited story.

The telescope did not merely create wonder. It created correction.

It forced human beings to compare belief with evidence.

That is why Galileo belongs so centrally in this book.

He showed that the sky could be examined, not merely worshiped. He helped shift humanity from reverence toward investigation.

And once investigation began, the old sacred heavens could not survive unchanged.

What Galileo Did Not Prove

It is important to be fair.

Galileo did not prove atheism. He did not disprove the existence of God. He did not show that religion has no emotional or cultural value. He did not solve all astronomical problems. He did not complete the scientific revolution by himself.

He also made mistakes. Like all human beings, he was a man of his time, with ambition, pride, courage, brilliance, and limitation.

But none of that weakens his importance.

His observations showed that inherited authority could be wrong about nature. They showed that direct evidence could challenge sacred and philosophical tradition. They showed that instruments could extend human perception beyond ancient limits. They showed that the sky was not the perfect divine realm many had imagined.

The importance of Galileo is not that he destroyed God.

The importance of Galileo is that he helped destroy the need for God-shaped astronomy.

After Galileo, the heavens could not easily remain a place of untouchable perfection. They had become rough, changing, measurable, and physical.

That was enough.

The Religious Reply

A thoughtful believer may respond:

"Galileo did not disprove God. The Moon can have mountains and still be created by God. Jupiter can have moons and still exist within divine creation. The Sun can have spots and still be part of God's universe. Galileo only corrected a mistaken interpretation of nature, not the existence of God."

This reply is reasonable.

It is true that Galileo did not disprove God. It is true that a religious person can accept telescopic discoveries. It is true that many believers today have no difficulty accepting mountains on the Moon, moons around Jupiter, phases of Venus, sunspots, and modern astronomy.

But this reply gives up more than it first appears to give up.

It admits that religious tradition was not the reliable authority on the physical structure of the heavens. It admits that observation corrected inherited belief. It admits that the older picture of perfect heavens was wrong.

The believer may preserve God by moving God away from the specific explanation. God no longer explains why the heavens appear as they do. Astronomy does that.

This is the recurring pattern.

Religion adapts after science explains.

Evidence-Based Response

The question is not whether belief in God can survive Galileo. It can.

The question is whether God remains necessary to explain the heavens Galileo observed.

The answer is no.

The mountains on the Moon require light, shadow, surface features, and physical explanation, not divine perfection.

The moons of Jupiter require orbital motion, not a universe centered only on Earth.

The phases of Venus require planetary geometry, not sacred symbolism.

The spots on the Sun require solar physics, not a flawless heavenly realm.

Galileo's discoveries helped move the heavens from theology to science.

They showed that nature could be investigated directly and that religiously comfortable ideas could be wrong. They also showed that the universe does not owe us emotional reassurance.

The heavens are not perfect because we want them to be.

Earth is not central because we feel important.

The sky is not sacred because ancient people imagined it so.

Reality must be allowed to speak for itself.

And when Galileo pointed his telescope upward, reality spoke with unusual force.

Chapter Summary

Galileo's telescope changed the human view of the heavens.

For centuries, many people imagined the heavens as perfect, pure, and fundamentally different from the changeable Earth. Galileo's observations challenged that belief. He saw mountains and valleys on the Moon, moons around Jupiter, phases of Venus, sunspots, and far more stars than the naked eye could see.

These discoveries did not disprove God, but they weakened the old sacred picture of the sky. The heavens became physical, observable, imperfect, and subject to investigation.

The Galileo story is not a simple tale of science defeating religion in every sense. But it is a powerful example of evidence challenging inherited authority.

After Galileo, the sky could no longer remain the untouched realm of perfect divine imagination.

Key Points

The ancient and medieval worldview often treated the heavens as perfect, pure, and closer to the divine.

Galileo's telescope revealed an imperfect and physical sky.

The Moon had mountains and valleys.

Jupiter had moons of its own.

Venus showed phases that supported the Copernican direction of thought.

Sunspots challenged the idea of a flawless Sun.

Galileo's conflict with religious authority showed the danger of allowing doctrine to control claims about nature.

His work did not disprove God, but it made God less necessary as an explanation of the heavens.

Transition to the Next Chapter

Galileo showed that the heavens were not perfect.

But science needed more than observation. It needed mathematical law.

The sky had to be not only seen differently, but understood differently. The planets did not wander randomly. They moved in patterns that could be described with precision.

That next step came through Johannes Kepler.

In the next chapter, we will see how the sky became mathematical.

Sources

Galileo Galilei, *Sidereus Nuncius*

Galileo Galilei, *Dialogue Concerning the Two Chief World Systems*

Stanford Encyclopedia of Philosophy, "Galileo Galilei"

Encyclopedia Britannica, "Galileo Galilei"

Encyclopedia Britannica, "Galileo: Telescopic Discoveries"

NASA, "Galileo's Observations of the Moon, Jupiter, Venus and the Sun"

John Hedley Brooke, *Science and Religion: Some Historical Perspectives*

Thomas S. Kuhn, *The Copernican Revolution*

Stillman Drake, *Galileo at Work*

Maurice A. Finocchiaro, *The Galileo Affair*

Jacob Bronowski, *The Ascent of Man*

Carl Sagan, *Cosmos*

Chapter 5: Kepler and the Sky Written in Mathematics

When the Planets Stopped Wandering

For most of human history, the planets looked mysterious.

The stars seemed fixed in their places, forming familiar patterns in the night sky. But the planets behaved differently. They moved. They wandered. Sometimes they seemed to slow down. Sometimes they seemed to stop. Sometimes they even appeared to move backward for a while before continuing forward again.

To ancient observers, this was strange and meaningful.

The word planet itself comes from the idea of a wanderer. The planets were the wandering lights of the heavens. Because their motion seemed unusual, many cultures gave them special importance. They were treated as signs, powers, divine beings, or heavenly messengers. Astrology grew from this ancient belief that the movements of the planets were connected to human fate.

If Mars appeared in a certain place, perhaps war was coming.

If Venus shone brightly, perhaps love or fertility was favored.

If Saturn moved through a certain region of the sky, perhaps hardship or misfortune would follow.

The sky was not only watched.

It was interpreted.

The planets seemed to carry hidden meaning.

Then came a different way of seeing them.

What if the planets were not messengers?

What if they were not signs of human destiny?

What if they were bodies moving according to mathematical law?

That question changed the sky.

Copernicus had moved Earth away from the center. Galileo had shown that the heavens were not perfect. But Johannes Kepler gave the moving sky something even more powerful: mathematical order.

The planets did not wander because they were divine signs.

They moved because nature followed law.

The Dream of Perfect Circles

Before Kepler, many thinkers believed that heavenly motion should be circular.

The circle seemed perfect. It had no beginning and no end. It was smooth, symmetrical, and complete. For minds shaped by Greek philosophy and later religious imagination, circular motion seemed appropriate for the heavens. If the heavens were higher and more perfect than Earth, then their motion should also be perfect.

This idea was beautiful.

It was also wrong.

The problem was that the planets did not move in simple perfect circles. Their observed positions did not fit that dream easily. Astronomers had to add complicated devices to preserve circular motion. They used epicycles, deferents, and other mathematical adjustments to explain why planets appeared where they did.

The old system became increasingly complex.

This is one of the dangers of protecting a beautiful idea after nature has begun to resist it. Instead of giving up the idea, people often build more defenses around it. They add explanation after explanation, adjustment after adjustment, until the system becomes heavy and strained.

Kepler inherited this problem.

He wanted order. He wanted harmony. He believed deeply that the universe was mathematically intelligible. He was not an atheist trying to remove God from the sky. Like many scientists of his time, he saw mathematics as a way of understanding divine creation.

But Kepler also had a rare intellectual honesty.

When the data did not fit the perfect circle, he did not force the heavens to obey an old ideal.

He allowed the circle to die.

That was a major moment in the history of science.

The circle was beautiful.

But the ellipse was true.

The Astrologer and the Astronomer

Kepler lived in a world where astrology still had influence.

Many people looked at the planets not only to understand the sky, but to understand themselves. They wanted to know whether Mars warned of war, whether Venus promised love, whether Saturn carried misfortune, whether Jupiter brought favor, and whether the stars had written something about their future.

This was not unusual in his time. The line between astronomy and astrology was not as clean as it is today. Kepler himself made astrological calendars and lived within a culture that still gave the heavens personal meaning.

But his greatest work helped weaken that very worldview.

This is one of history's ironies.

The more carefully the planets were studied, the less they looked like messages about human destiny. Their motion was not a private code written for kings, lovers, armies, or farmers. Their paths could be described by mathematics.

Mars was not angry.

Venus was not arranging love.

Saturn was not punishing anyone.

Jupiter was not announcing royal favor.

They were planets.

Bodies in motion.

The sky was moving from astrology to astronomy.

From omen to orbit.

From personal meaning to natural law.

Tycho's Data and Kepler's Struggle

Kepler's breakthrough depended on the careful observations of Tycho Brahe.

Tycho was one of the greatest observational astronomers before the telescope. He measured planetary positions with remarkable precision for his time. His data were far better than what earlier astronomers had possessed.

Kepler eventually gained access to these observations, especially the detailed records of Mars.

Mars became the problem.

Its motion was difficult to explain with perfect circles. Kepler worked intensely to fit the data. He tried old assumptions. He tested different arrangements. He struggled with the numbers. He refused to accept a solution that was merely close enough if the observations said otherwise.

This is one of the most important features of science.

The data mattered more than the preferred idea.

Kepler could have protected the ancient circle. He could have treated the mismatch as minor. He could have forced the evidence into inherited geometry. But he did something more courageous.

He followed the evidence.

He found that Mars did not move in a perfect circle around the Sun. Its path was an ellipse, with the Sun at one focus.

This discovery was more than a technical correction. It was a symbolic defeat for heavenly perfection. The planets moved beautifully, but not according to the old human dream of perfection. They moved according to the real geometry of nature.

Kepler had discovered that the heavens were lawful, but not in the way people had imagined.

The First Law: The Circle Gives Way

Kepler's first law says that planets move in elliptical orbits, with the Sun at one focus.

This may sound simple now, but it was revolutionary.

An ellipse is not a broken circle. It is not an imperfect mistake. It is a precise mathematical shape. The planets were not wandering randomly. They were not drifting under mysterious spiritual influence. They followed a lawful path.

But the path was not the perfect circle that philosophers had loved.

This was another blow to the old sacred sky.

Galileo had shown that the Moon was rough and the Sun was spotted. Kepler showed that planetary motion itself did not obey the ancient demand for circular perfection. Nature was mathematical, but it was not obligated to match human ideals of beauty.

This is a recurring lesson in science.

Nature is not required to satisfy our imagination.

We may prefer a perfect circle, but the planet follows an ellipse. We may prefer a central Earth, but Earth moves around the Sun. We may prefer a young Earth, but geology reveals deep time. We may prefer special creation, but biology reveals evolution. We may prefer the soul to float free from matter, but the mind depends deeply on the brain.

Reality has its own structure.

Science becomes powerful when it allows reality to correct our preferences.

Kepler's first law was an act of surrender to reality.

The Second Law: Motion Has Pattern

Kepler's second law showed that a planet does not move at the same speed throughout its orbit.

When a planet is closer to the Sun, it moves faster. When it is farther away, it moves more slowly. More technically, a line connecting the planet to the Sun sweeps out equal areas in equal times.

The details are less important here than the meaning.

Planetary motion was not chaotic. It was not arbitrary. It was not a series of divine impulses. It had a pattern that could be described mathematically.

This was another step away from mystery.

In the older religious imagination, unusual motion invited supernatural interpretation. A strange movement in the sky could be a sign. A wandering planet could carry meaning. A celestial event could be connected to fate, blessing, or judgment.

Kepler's work pointed in a different direction.

The planet moves faster here and slower there because its motion follows a lawful relationship. The explanation is not moral. It is mathematical.

This shift is profound.

Once motion becomes mathematical, it no longer needs to be read as a message. It can be studied as a pattern. It can be calculated. It can be predicted.

Prediction reduces fear.

If we can predict an eclipse, it becomes less terrifying. If we can calculate planetary motion, the planets become less mysterious. If we can understand the sky, we are less tempted to read every light in heaven as a sign about human affairs.

Mathematics did not make the sky less beautiful.

It made the sky less superstitious.

The Third Law: Harmony Without Mysticism

Kepler's third law connected the time a planet takes to orbit the Sun with its distance from the Sun.

Again, the mathematical details are not the main point of this book. The larger meaning is this: the solar system had deep order. The planets were not separate mysteries. Their motions were connected by a general mathematical relationship.

Kepler called one of his works *The Harmony of the World*. He believed he was discovering a deep harmony in creation. His language could be religious, mystical, and mathematical all at once. He was a complex figure, not a modern secular scientist in the simple sense.

This matters because we must be honest.

Kepler did not remove God from his own imagination. He often saw mathematical order as evidence of divine reason. But the effect of his science was still to reduce the need for direct divine explanation.

The planets did not need to be guided individually by spiritual forces. They did not need to be interpreted as omens. They did not need to be understood through astrology. Their motion could be described by law.

Kepler's own religious belief did not change the scientific meaning of his discovery.

This is a pattern we will see again and again.

A scientist may believe in God. But if his discovery explains a natural process without invoking God, then that discovery reduces the explanatory need for God.

The person may remain religious.

The explanation becomes natural.

Law Replaces Sign

One of the great changes in the history of science was the movement from sign to law.

In the older worldview, the sky often functioned as a field of signs. Celestial events were read for meaning. People looked upward not only to see what was happening, but to discover what heaven intended.

A comet could warn.

An eclipse could threaten.

A planet could influence.

A star could announce.

A strange light could predict disaster.

The heavens were like a sacred text.

Kepler's work helped move the sky away from this kind of interpretation. The planets were not signs in a divine message system. They were bodies moving according to mathematical law.

This did not happen all at once. Astrology did not vanish overnight. Superstitious readings of the sky continued, and still continue today in different forms. But intellectually, a major shift had occurred.

The sky became less like scripture and more like mathematics.

This was a decisive change.

A sign depends on interpretation. It can be stretched, adapted, and reinterpreted. A law must be tested against reality. It must describe what happens. It must predict. It must survive comparison with observation.

The movement from sign to law made science stronger than divination.

It also made God less necessary as an explanation.

If Mars moves according to mathematical law, we do not need to ask what Mars means for the fate of a king. If Venus shows a calculable pattern, we do not need to ask what goddess is speaking through it. If planetary motion follows law, we do not need celestial spirits to guide it.

The planets were no longer divine messengers.

They were natural bodies.

Mathematics as a New Language of Reality

Kepler's work revealed something astonishing: nature could be written in mathematics.

This was one of the deepest discoveries in human history.

Mathematics is not a prayer. It is not a ritual. It is not a myth. It does not flatter human emotion. It does not care about social rank, priestly authority, or inherited belief. It is precise, disciplined, and demanding.

When mathematics describes nature successfully, it gives human beings a new kind of power.

It allows prediction.

It reveals hidden order.

It exposes error.

It connects separate events.

It allows one mind to check another.

It lets future generations build on past discovery.

The old religious explanation often said, “The gods move the heavens.”

Mathematics asked, “By what pattern?”

That question was devastating to supernatural explanation.

If the pattern can be found, measured, and used to predict future motion, then the supernatural claim becomes unnecessary. The god may remain as a belief, but not as the working explanation.

This is why Kepler belongs in this book.

He did not simply improve astronomy. He helped reveal that the sky was governed by law. He helped transform the heavens from sacred mystery into mathematical reality.

The more mathematical the sky became, the less it needed mythology.

Natural Explanation Proves Itself

Kepler's laws did not merely sound elegant.

They worked.

They described planetary motion better than the older insistence on perfect circles. They organized observations into a clearer pattern. They allowed later thinkers to ask deeper questions. They prepared the way for Newton's laws of motion and universal gravitation.

This practical success matters.

A myth may make the planets meaningful.

Astrology may make them personal.

A sacred story may make them morally important.

But Kepler's mathematics made them understandable.

That is the difference.

A person can believe Mars warns of war, but that belief does not accurately calculate the orbit of Mars. A person can believe Venus governs love, but that belief does not explain the phases or motion of Venus. A person can believe the heavens are signs from God, but that belief does not reveal the mathematical relationship between orbital period and distance from the Sun.

Kepler's laws did explanatory work.

They did not simply decorate the sky with meaning.

They showed how the sky behaved.

When an explanation works better, predicts better, and opens the way to deeper knowledge, it earns authority. Not because it is sacred. Not because it is ancient. Not because it comforts us. Because it answers to reality.

That is why science replaced the old reading of the sky.

The Decline of Astrology

Kepler lived at a time when astrology was still widely taken seriously. He himself made astrological calendars and did not entirely separate astronomy from astrology in the modern way. This may surprise modern readers.

But history is rarely clean.

People do not suddenly step from darkness into full light. They carry old beliefs while discovering new truths. Kepler could still use astrological language while also developing the mathematics that would help destroy astrology's intellectual foundation.

The same careful study of planetary motion that once served astrology eventually weakened it.

Astrology depends on the idea that celestial bodies carry personal meaning for human lives. It assumes a link between planetary positions and human character, fate, love, war, fortune, or suffering. But as astronomy matured, planets became physical bodies moving through space according to law. Their motion could be calculated without reference to human destiny.

Mars was not angry.

Venus was not arranging love.

Saturn was not imposing sorrow.

Jupiter was not granting royal favor.

They were planets.

That realization did not remove all human fascination with astrology. Many people still enjoy or believe in it. But scientifically, astrology lost its foundation. It could not compete with astronomy because it did not offer reliable evidence, mechanism, or predictive success in the way science requires.

Kepler's work helped the sky move from omen to orbit.

That was a major step in making God and the gods unnecessary as explanations of celestial motion.

Kepler and the Religious Mind

Kepler was not an enemy of religion.

He lived in a deeply religious age. He used religious language. He saw mathematical order as part of God's creation. He believed the human mind could understand the universe because both mind and universe were connected to divine reason.

This should be stated clearly.

It would be wrong to pretend that Kepler was a modern atheist. He was not.

But this fact does not weaken the argument of this chapter. It strengthens it, because it shows the argument is not about the private belief of the scientist. It is about the public power of the explanation.

Kepler's faith may have inspired his search for order. But his scientific achievement did not depend on saying, "God moves the planets." It depended on observation, data, geometry, and mathematical law.

That is the key distinction.

Religion may inspire a person. Science explains the process.

A religious scientist can make a discovery that reduces the need for religious explanation. This happened many times in history.

Kepler's God could remain in his heart and theology. But the planets themselves increasingly belonged to mathematics.

That is the retreat of God in its more subtle form.

God is not always pushed out by direct denial. Sometimes God is displaced by a better explanation, even from the hands of a believer.

The Religious Reply

A thoughtful believer may say:

"Kepler's discoveries do not remove God. They reveal the beautiful mathematical order that God placed in the universe. Natural laws are not enemies of God. They are the way God governs creation. Kepler himself believed that mathematical harmony reflected divine wisdom."

This reply is historically fair.

Kepler did see mathematics in religious terms. Many believers have understood natural law as the regularity of divine creation. For them, scientific law does not remove God. It reveals God's method.

This is a serious position.

But once again, it changes God's role.

God is no longer needed to explain the specific motion of Mars, Venus, Jupiter, or Earth. The law explains the motion. The believer may place God behind the law, but that is a broader theological claim, not a necessary astronomical explanation.

Science can calculate the orbit without invoking God. It can test the law without prayer. It can correct the model without scripture. It can predict planetary motion without asking what the heavens mean spiritually.

That is the important point.

God may remain as an interpretation of why there is order. But God is no longer required to describe how the planets move.

Evidence-Based Response

Kepler's work made the sky more lawful and less mystical.

His laws showed that planetary motion could be described with mathematical precision. The planets were not wandering signs. They were bodies following regular paths. Their motion did not require astrology, divine messages, angelic movers, or sacred symbolism.

The old idea of perfect circular motion gave way to the more accurate ellipse. This mattered because it showed that nature must correct human ideals, not the other way around.

Kepler did not disprove God.

But he did weaken the need for God as an explanation of planetary motion.

That is exactly the pattern this book follows.

Where a natural law explains the evidence, divine explanation becomes unnecessary.

The sky remained beautiful. It remained majestic. It remained capable of inspiring awe. But after Kepler, its beauty was no longer only the beauty of sacred mystery. It was the beauty of mathematical order.

And mathematical order could be studied without theology.

Chapter Summary

Kepler helped transform the sky from a realm of mystical wandering into a realm of mathematical law.

Before him, the planets were often seen as wandering lights with possible divine or astrological meaning. The heavens were expected to move in perfect circles because circles seemed pure and fitting for the celestial realm.

Using Tycho Brahe's precise observations, especially of Mars, Kepler discovered that planets move in elliptical orbits. He showed that planetary motion followed mathematical relationships. The sky was not random, but it also did not obey older human ideals of divine perfection.

Kepler himself was religious, but his science reduced the need for direct divine explanation. The planets no longer needed to be understood as omens, divine signs, or spiritually guided wanderers. They could be understood as natural bodies moving according to law.

Key Points

The planets once seemed mysterious because they appeared to wander through the sky.

Ancient and medieval thinkers often expected heavenly motion to be circular and perfect.

Kepler used careful observations to show that planets move in ellipses, not perfect circles.

His work helped replace celestial mystery with mathematical law.

Kepler was personally religious, but his discoveries made God less necessary as an explanation of planetary motion.

The sky became less like a sacred message and more like a lawful natural system.

Transition to the Next Chapter

Kepler gave the planets mathematical paths.

But one great question remained.

Why did they move that way?

What force connected falling objects on Earth with the motion of the Moon and planets in the sky? What law could unite the apple, the Moon, the Earth, and the Sun?

That next step belonged to Isaac Newton.

In the next chapter, we will see how Newton brought Earth and heaven under one law.

Sources

Johannes Kepler, *Astronomia Nova*

Johannes Kepler, *Harmonices Mundi*

Johannes Kepler, *Epitome of Copernican Astronomy*

Max Caspar, *Kepler*

Arthur Koestler, *The Sleepwalkers*

Thomas S. Kuhn, *The Copernican Revolution*

Owen Gingerich, *The Eye of Heaven*

James R. Voelkel, *Johannes Kepler and the New Astronomy*

Encyclopedia Britannica, "Johannes Kepler"

Stanford Encyclopedia of Philosophy, "Johannes Kepler"

Carl Sagan, *Cosmos*

Jacob Bronowski, *The Ascent of Man*

John Hedley Brooke, *Science and Religion: Some Historical Perspectives*

Chapter 6: Newton and the Universe Under One Law

The Apple and the Moon

A falling apple is an ordinary thing.

A child can see it. A farmer can see it. Anyone sitting under a tree can see it. The apple loosens from the branch, drops through the air, and lands on the ground. Nothing could seem more common.

The Moon, by contrast, feels distant and mysterious. It shines in the night. It changes shape. It moves silently across the sky. For most of human history, the Moon belonged to the heavens, to poetry, to religion, to calendar, to myth, and to imagination.

An apple falls in a garden.

The Moon travels through the sky.

Before modern science, these seemed like different kinds of events. One belonged to Earth. The other belonged to heaven. One happened in the world of soil, trees, hunger, weather, and daily labor. The other happened in the high realm above, where people imagined divine order, perfection, and mystery.

Isaac Newton saw something deeper.

He asked whether the same force that pulls the apple toward Earth might also help hold the Moon in its path.

That question changed the universe.

It connected the ordinary with the cosmic. It joined the garden and the sky. It suggested that Earth and heaven were not governed by separate rules. The same natural law could explain the fall of an apple and the motion of the Moon.

This was one of the greatest acts of unification in human thought.

The apple and the Moon were connected.

Earth and heaven came under one law.

The Old Divide Between Earth and Heaven

For centuries, many educated thinkers inherited the idea that the heavens were fundamentally different from the Earth.

Earth was the realm of change. Things here were heavy, unstable, and corruptible. Bodies grew, weakened, broke, burned, rusted, and died. Stones fell. Smoke rose. Flesh decayed. Storms came and went. Nothing remained perfect for long.

The heavens seemed different.

The stars appeared fixed. The planets moved in regular patterns. The Moon returned again and again. The Sun rose and set with dependable majesty. Compared with the human world below, the sky looked orderly, permanent, and pure.

This division supported religious imagination. Heaven was above. God was above. Angels, divine order, and spiritual perfection were imagined above. Earth was below, the place of sin, suffering, labor, and death.

The old universe had hierarchy.

Higher meant better.

Lower meant lesser.

Heaven meant perfection.

Earth meant corruption.

Copernicus had already disturbed this structure by moving Earth away from the center. Galileo had damaged it further by showing mountains on the Moon, moons around Jupiter, and spots on the Sun. Kepler had shown that planets followed elliptical paths rather than perfect circles.

But Newton went even deeper.

He did not merely say that the heavens were imperfect. He showed that the heavens and the Earth obeyed the same mathematical law.

This was devastating to the old division.

If one law governs both the falling apple and the orbiting Moon, then the heavens are not a separate divine realm exempt from natural explanation. They are part of the same physical universe.

The sacred ceiling had become a lawful sky.

Newton's Great Synthesis

Newton's great work, *Mathematical Principles of Natural Philosophy*, usually called the *Principia*, was published in 1687. It is one of the most important books in the history of science.

Newton did not appear from nowhere.

Copernicus had moved Earth. Galileo had studied motion and falling bodies. Kepler had described the mathematical paths of planets. Others had contributed to mechanics, astronomy, optics, and mathematics. Science is rarely the work of one isolated genius. It is usually a long conversation across generations.

But Newton's achievement was extraordinary.

He brought together motion, force, gravity, mathematics, and astronomy in a new and powerful way. He gave laws of motion. He described universal gravitation. He showed that the same principles could apply to objects on Earth and bodies in the heavens.

This is why Newton was so revolutionary.

He did not simply add another fact.

He created a new kind of universe.

A universe of law.

A universe of mathematics.

A universe of prediction.

A universe in which planets, projectiles, tides, falling objects, and celestial paths could be studied through the same principles.

Before Newton, the sky still carried much mystery. After Newton, the sky became calculable in a new way.

This did not remove wonder.

It changed the nature of wonder.

The old wonder asked, “What divine power moves the heavens?”

The new wonder asked, “How can so much of nature be understood through law?”

Universal Gravitation

Newton’s law of universal gravitation was powerful because of one word: universal.

Gravity was not only a local tendency of heavy things to fall downward. It was not only something that happened near Earth. It was a general attraction between masses.

Every object with mass attracts every other object with mass.

The Earth attracts the apple. The apple also attracts the Earth, though the effect on Earth is tiny because Earth is so massive. The Earth attracts the Moon. The Sun attracts the planets. The planets attract one another.

This was a profound idea.

It connected all physical bodies into one system of interaction. The universe was not a collection of separate sacred zones. It was a unified natural order.

The same law applied everywhere.

That is why Newton weakened supernatural explanation so deeply. The heavens no longer needed special rules. Planetary motion did not require divine pushing, angelic guidance, or separate celestial perfection. The Moon did not remain in orbit because it belonged to a higher spiritual realm. It remained in orbit because of motion and gravity.

Newton made the sky physical in a deeper way than Galileo had.

Galileo showed that the heavens had physical features.

Newton showed that the heavens had physical causes.

That difference matters.

To see mountains on the Moon is powerful. To explain the Moon's motion by the same principles that explain falling bodies is even more powerful.

It brings the Moon down into physics.

Motion Without Divine Pushing

One old problem was motion.

Why do things move? Why do they continue moving? Why do planets move in their paths? Why does the Moon not fall straight down to Earth? Why does Earth not fly away from the Sun?

In older ways of thinking, motion often seemed to need a mover. Something had to push, guide, direct, or sustain movement. This made it easy to imagine divine action behind motion. If the planets moved, perhaps spiritual beings moved them. If the heavens turned, perhaps divine intelligence turned them.

Newton changed this picture.

His laws of motion showed that objects behave in regular ways. A body at rest remains at rest, and a body in motion continues in motion unless acted upon by a force. Force changes motion. Action and reaction are related. Gravity acts between masses.

These ideas allowed motion to be explained without constant supernatural intervention.

A planet does not need to be pushed around the Sun by an angel. Its motion can be understood as the combination of its forward motion and the gravitational attraction of the Sun. The Moon does not need to be held in place by divine command. Its orbit can be understood through motion and gravity.

This was a major retreat of divine explanation.

God had once been imagined as directly moving or sustaining the heavens. Newtonian physics made that unnecessary as a working explanation. The universe could be described as a system of bodies moving according to law.

This did not prove there was no God.

But it made God less necessary in physics.

Prediction and Power

Newton's laws were not merely beautiful.

They worked.

That is what made them so powerful.

They allowed scientists to calculate motion with remarkable success. They helped explain tides, planetary orbits, comets, falling bodies, projectiles, and many other physical events. They gave human beings the ability to predict and understand the behavior of objects in ways that older explanations could not.

Prediction is one of the great tests of explanation.

A myth may comfort.

A sacred story may inspire.

A ritual may unite a community.

But a scientific law can predict.

This matters.

If a theory can predict the return of a comet, the motion of a planet, the path of a projectile, or the effect of force, then it has a kind of strength that divine explanation rarely has.

"God moves the planets" does not tell us where Mars will be next month.

Newtonian mechanics can calculate.

"God sends the tide" does not explain its timing and relation to the Moon and Sun.

Newtonian gravity can clarify.

"God holds the heavens in order" may be emotionally meaningful to a believer, but it does not give the mathematical structure of motion.

Science did.

This is why natural law gradually replaced divine explanation. It did not merely deny. It explained. It predicted. It connected. It worked.

The Practical Power of a Lawful Universe

Newton's physics did not remain locked in books.

It entered the world.

It helped human beings understand motion, machines, engineering, navigation, artillery, tides, bridges, and eventually space travel. Later science would go far beyond Newton, but Newton's laws still remain powerful for ordinary motion and many practical purposes.

This matters for the argument of this book.

Natural explanation proves itself by consequences.

When a bridge stands because engineers understand forces, that is natural law at work.

When a ship navigates more accurately because astronomy and mathematics are reliable, that is natural explanation at work.

When a satellite is placed into orbit, it does not stay there because of prayer, ritual, or sacred text. It stays there because gravity, motion, velocity, mass, and calculation have been understood well enough.

This is the difference between explanation and reverence.

A person may look at the Moon and feel reverence. That is human and beautiful. But reverence does not calculate an orbit. Reverence does not predict a tide. Reverence does not send a spacecraft.

Physics does that.

Natural law did not merely change what humans believed.

It changed what humans could do.

That practical success made the old supernatural explanation weaker, not by insult, but by results.

Newton Believed in God

It is important to be honest about Newton.

Newton was not an atheist. He believed in God deeply, though his private theology was complex and not fully orthodox by later Christian standards. He studied scripture, prophecy, chronology, and theology intensely. He did not see himself as destroying religion. He believed that the order of nature reflected divine intelligence.

This must be acknowledged clearly.

The argument of this book does not depend on pretending that every great scientist was an atheist. That would be false. Many were religious. Many saw their work as a way of understanding God's creation.

Newton is one of the clearest examples.

But this does not weaken the argument. It makes the argument more precise.

Newton's personal belief in God is not the same as the explanatory role of God in Newtonian physics.

In his science, Newton did not explain planetary motion by saying, "God pushes the planets." He used laws of motion and gravitation. His religious belief may have shaped his sense that the universe was orderly, but the scientific explanation itself worked through natural law.

This distinction is essential.

A believer can discover a natural explanation that makes God unnecessary for that specific phenomenon.

That is exactly what happened many times in the history of science.

Newton may have believed in God, but after Newton, the heavens needed less divine management.

A Lawful Universe and the Distant God

Newton's universe was so lawful that it created a new problem for theology.

If the universe runs by stable laws, what exactly is God doing?

In older religious imagination, God could be placed directly behind events. God sent rain. God moved the heavens. God punished through plague. God directed storms. God's hand was seen everywhere.

But a lawful universe changes that.

If rain follows atmospheric conditions, if disease follows biological processes, if planets follow gravity and motion, if tides follow the Moon and Sun, then God becomes less visible as a direct cause.

Some religious thinkers adapted by imagining God as the creator of the laws rather than the constant direct cause of each event. God became the divine lawgiver, the architect, the designer of the system.

This view preserved belief, but it also moved God farther away from daily explanation.

God was no longer needed to explain why the Moon orbits Earth. Gravity did that. God was no longer needed to explain why planets move in predictable paths. Mechanics did that. God was no longer needed to explain why projectiles curve. Physics did that.

God became more abstract.

This is another form of retreat.

Religion did not vanish. But its explanatory territory shrank.

The more lawful the universe became, the less room there was for God as an immediate physical explanation.

The Clockwork Image

After Newton, many thinkers began to imagine the universe as a kind of machine.

This image is often called the clockwork universe.

A clock has parts. The parts fit together. The gears move according to mechanical principles. Once understood, the clock does not need spirits inside it to make each gear turn. It works through structure and law.

The Newtonian universe suggested something similar. Nature seemed ordered, lawful, and intelligible. The motions of bodies could be calculated. The same principles applied across vast distances. The universe appeared less like a magical drama and more like a physical system.

For some believers, this strengthened belief in a creator. If there is a clock, there must be a clockmaker, they argued.

But the clockwork image also weakened the need for ongoing divine intervention. Once the machine is understood, one does not need to invoke God to explain each movement of the gears.

This was not atheism, but it was a step toward a more natural universe.

The God of direct action became less necessary. The God of background design remained possible, but less involved in daily explanation.

This is a crucial movement in the history of thought.

God moved from the storm to the law behind weather.

From the orbit to the law behind motion.

From the disease to the laws of biology.

From the object to the system.

From immediate cause to distant cause.

Each move preserved belief for some, but reduced God's explanatory role.

One Law for Earth and Heaven

The greatness of Newton lies partly in unification.

Human beings often divide reality into compartments. Sacred and ordinary. Earthly and heavenly. Natural and supernatural. Human and divine. Body and soul. Matter and spirit.

Science often progresses by showing that things once separated are connected.

Newton connected Earth and heaven.

The falling apple and the orbiting Moon were not different kinds of events in different cosmic realms. They were part of one lawful system.

This unification had deep consequences.

If the same law applies to Earth and heaven, then heaven is not scientifically separate from Earth. It may remain symbolically or religiously important to believers, but as a physical region, it belongs to the same universe.

The heavens are not exempt.

They can be measured.

They can be calculated.

They can be predicted.

They can be explained.

They can be brought under natural law.

This is one of the major steps in making God unnecessary.

A separate sacred sky needs divine explanation. A lawful physical sky does not.

The universe became less divided, less magical, and less centered on religious imagination.

It became more coherent.

It became more natural.

The Human Meaning of Newton's Discovery

Newton's discovery changed more than physics. It changed how human beings could think about reality.

If the universe obeys laws, then events are not merely expressions of divine mood. If nature has regularity, then human beings can investigate it. If motion can be described mathematically, then knowledge can grow beyond myth. If the same laws apply everywhere, then the universe is open to reason.

This gave humanity extraordinary confidence.

Not arrogance, though arrogance sometimes followed. The deeper confidence was this: reality can be understood.

That idea is powerful.

A world ruled by unpredictable divine will creates fear. A world governed by natural law creates possibility. We may not control everything, but we can learn. We can predict. We can build. We can navigate. We can engineer. We can explore.

Newton helped make the universe intellectually accessible.

He did not make it small. He did not remove mystery entirely. But he showed that the same rational mind that studies a falling body can also understand celestial motion.

The sky was no longer beyond human explanation.

That was a great blow to supernatural thinking.

It was also a great victory for human reason.

The Limits of Newton

Newton's physics was not the final word.

Later, Einstein would transform our understanding of gravity, space, and time. Quantum physics would challenge classical ideas at very small scales. Modern cosmology would reveal a universe more vast and dynamic than Newton could have imagined.

Science moved beyond Newton.

But this does not reduce his importance. It actually reveals the strength of science.

Newton was corrected and expanded, not worshiped as untouchable scripture. His work remained immensely useful in many areas, but later evidence and theory showed its limits. Science could honor Newton without freezing thought forever at Newton.

This is another difference between science and sacred authority.

A great scientist can be surpassed.

A scientific theory can be revised.

A powerful explanation can remain useful within limits while being replaced by a deeper explanation.

That is not weakness.

That is growth.

Newton's universe was not final, but it was a monumental step away from divine explanation. It showed that nature could be lawful on a scale human beings had scarcely imagined.

And once that door opened, science continued walking through it.

The Religious Reply

A thoughtful believer may say:

"Newton himself believed in God. He saw the order of the universe as evidence of divine intelligence. Natural laws do not remove God. They show the wisdom of the lawgiver. Gravity may describe how the planets move, but God created gravity and sustains the universe."

This reply deserves respect.

Newton did believe in God. Many religious thinkers have seen natural law as compatible with divine creation. It is reasonable to say that discovering a law does not logically disprove every possible creator behind the law.

But this reply also changes the role of God.

God is no longer needed to explain the motion itself. Gravity and motion explain that. God is moved behind the law as a theological interpretation.

That is not the same as scientific necessity.

A physicist can calculate planetary motion without invoking God. An engineer can use mechanics without theology. An astronomer can predict celestial events without prayer. The law does the explanatory work.

The believer may still say, "God made the law."

But science does not require that claim in order to function.

Evidence-Based Response

Newton's achievement did not prove atheism.

But it made a major part of divine explanation unnecessary.

Before Newton, the heavens still carried an aura of mystery and special status. After Newton, the same kinds of laws applied to Earth and sky. Falling bodies, orbiting moons, moving planets, tides, and celestial paths could be understood through natural principles.

This changed the intellectual landscape.

God could still be believed in. But God was less needed to explain physical motion. The heavens no longer required a separate divine physics. They belonged to the same lawful universe as Earth.

This is the historical pattern again.

Science does not need to disprove God in every philosophical sense. It only needs to explain the phenomenon better.

Newton explained motion better.

Therefore, in that field, God became unnecessary.

Chapter Summary

Newton brought Earth and heaven under one law.

Before Newton, many people imagined the heavens as fundamentally different from the Earth. The sky seemed higher, purer, and closer to the divine. Newton showed that the same principles governing falling objects on Earth also help explain the motion of the Moon, planets, and celestial bodies.

His laws of motion and universal gravitation created a new vision of the universe as lawful, mathematical, and unified.

Newton personally believed in God, but his science reduced the need for direct divine explanation in physics. The planets did not need angels to move them. The Moon did not need divine command to remain in orbit. The heavens did not require separate sacred rules.

After Newton, the universe looked less like a supernatural drama and more like a lawful natural system.

Key Points

Newton unified earthly and heavenly motion under the same natural laws.

Universal gravitation connected falling objects, the Moon, planets, and the Sun.

Newton believed in God, but his scientific explanations worked through natural law.

His physics reduced the need for direct divine intervention in the heavens.

The idea of a lawful universe moved God from immediate cause to distant theological interpretation.

Newton did not end mystery, but he made the physical universe far more explainable without God.

Transition to the Next Chapter

Newton showed that the heavens obeyed law.

But an important question remained.

If mathematics and natural law could explain the motion of the planets so well, was God needed at all as a working scientific hypothesis?

That question would later be sharpened by Pierre-Simon Laplace.

In the next chapter, we will examine the famous idea of the unneeded hypothesis.

Sources

Isaac Newton, *Mathematical Principles of Natural Philosophy*

Isaac Newton, *Opticks*

I. Bernard Cohen, *The Newtonian Revolution*

Richard S. Westfall, *Never at Rest: A Biography of Isaac Newton*

James Gleick, *Isaac Newton*

Encyclopedia Britannica, "Isaac Newton"

Stanford Encyclopedia of Philosophy, "Newton's Philosophy"

Stanford Encyclopedia of Philosophy, "Newton's Views on Space, Time, and Motion"

Alexandre Koyré, *From the Closed World to the Infinite Universe*

Jacob Bronowski, *The Ascent of Man*

Carl Sagan, *Cosmos*

John Hedley Brooke, *Science and Religion: Some Historical Perspectives*

Peter Harrison, *The Territories of Science and Religion*

Chapter 7: Laplace and the Unneeded Hypothesis

The Missing God in the Equation

There is a famous story about Pierre-Simon Laplace.

Laplace, the great French mathematician and astronomer, had written an important work on celestial mechanics. He presented a copy to Napoleon. According to the well-known account, Napoleon noticed that the book explained the motions of the heavens without mentioning God.

Napoleon is said to have asked why.

Laplace reportedly answered:

"I had no need of that hypothesis."

The exact wording and historical details have been discussed by scholars, but the meaning of the story has become larger than the conversation itself.

It captures one of the most important ideas in the history of science.

God was not needed in the calculation.

This was not a childish insult. It was not merely an atheist joke. It was a statement about scientific explanation. If the motion of the planets can be explained by mathematics, gravity, and natural law, then God is not needed as part of that explanation.

A scientist may personally believe in God. A philosopher may argue for God. A theologian may place God behind the laws of nature. But in the working model of celestial mechanics, God does no necessary work.

That is the point.

Laplace did not need God in the equation because the equation already explained the motion.

This chapter is about that powerful moment in human thought: the moment when God becomes not disproved, but unnecessary.

After Newton

Newton had shown that Earth and heaven could be brought under one law.

The falling apple and the orbiting Moon belonged to the same physical universe. Gravity, motion, and mathematics could explain the behavior of bodies on Earth and in the sky.

But Newton's own view of God remained active. Newton believed that God had created and sustained the universe. He also seemed to think that divine action might sometimes be needed to maintain the stability of the solar system. The universe was lawful, but God was still close to the machinery.

Later thinkers moved further.

They asked whether the system needed that kind of divine supervision at all.

Laplace took the mathematical universe seriously. He extended Newtonian mechanics with extraordinary skill. He studied planetary motion, probability, tides, the stability of the solar system, and many problems in physics and astronomy. His work helped show that celestial motion could be explained in a more complete and self-contained way through natural law.

The God who had once moved the heavens had already been weakened by Copernicus, Galileo, Kepler, and Newton.

Laplace made the point sharper.

If the laws are enough, why add God?

What Is a Hypothesis?

To understand Laplace's importance, we must understand the word hypothesis.

A hypothesis is not just any belief. In science, a hypothesis is an idea proposed to explain something. It should help us understand the evidence. It should clarify what is happening. It should guide investigation. Ideally, it should lead to predictions or observations that can be checked.

A useful scientific hypothesis does work.

It explains.

It predicts.

It connects facts.

It can be tested.

It can be corrected.

It can be replaced if it fails.

God does not function this way in celestial mechanics.

If we say, "God moves the planets," what exactly have we explained? We have named a powerful agent, but we have not described the mechanism. We have not calculated the orbit. We have not predicted the position of Mars. We have not explained why planets move faster at some points and slower at others. We have not shown how small gravitational effects between planets alter their paths. We have not improved the mathematics.

By contrast, natural law does explain these things.

Gravity, motion, mass, distance, and mathematics allow calculation. They connect one event to another. They make prediction possible. They let one scientist check another scientist's work. They allow error to be found and corrected.

That is why God becomes unnecessary as a scientific hypothesis.

Not because God has been disproved in every philosophical sense.

But because God adds nothing necessary to the working explanation.

Naming Is Not Explaining

One of the most common errors in religious explanation is the confusion between naming and explaining.

If lightning strikes and someone says, “God did it,” that names a supposed cause. But it does not explain lightning.

If disease spreads and someone says, “God is punishing us,” that gives a religious interpretation. But it does not explain infection.

If planets move and someone says, “God guides them,” that gives a theological answer. But it does not explain celestial mechanics.

A true explanation must do more.

It must show how something happens. It must connect cause and effect. It must help us understand why this event happens in this way and not some other way. It must make the world clearer.

“God did it” is too easy. It can be used for anything.

Why did the child recover? God did it.

Why did the child die? God did it.

Why did the rain come? God did it.

Why did the drought continue? God did it.

Why did the planet move? God did it.

Why did the earthquake destroy one city and spare another? God did it.

An answer that fits every possible outcome explains no specific outcome.

Science became powerful because it demanded more. It asked for mechanisms, patterns, measurements, predictions, and correction.

Laplace's statement matters because it refuses to confuse a religious label with a scientific explanation.

If the natural explanation works, adding God does not improve it.

A Universe That Could Run Without Supervision

Laplace helped strengthen the image of a universe governed by law.

In such a universe, the planets do not need constant divine adjustment. Their motions may be complex, but complexity is not the same as miracle. A difficult calculation is not a gap for God. A system too large for ordinary intuition is not evidence of supernatural action.

This was an important change.

In earlier thought, the order of the universe often seemed to require an active divine manager. The heavens moved because God or divine intelligences sustained them. The stability of the cosmos depended on divine care. Without God's direct involvement, the system might seem to fall apart.

Laplace's work pointed in another direction.

The system could be studied as a natural system.

Its future motions could, in principle, be understood from its present state and the laws governing it. The universe did not look like a kingdom requiring constant royal command. It looked more like a lawful order whose behavior followed from natural principles.

This did not remove all mystery. It did not explain why there is a universe at all. It did not answer every philosophical question. But it did remove a large part of the old need for divine supervision.

The planets did not need God as an astronomer.

They moved according to law.

The Power and Limits of Determinism

Laplace is also associated with a strong form of determinism.

In a famous idea often called "Laplace's demon," he imagined an intelligence that knew the position and motion of every particle in the universe at one moment and knew all the laws of nature. Such an intelligence, he suggested, could know the future and the past with complete precision.

Modern science has changed this picture. Quantum physics, chaos theory, and the limits of measurement have made simple determinism less secure than it once seemed. We should not present Laplace as if he gave the final view of reality. He did not.

But his deeper historical importance remains.

Laplace represented a universe increasingly understood through natural law rather than divine will. Whether or not strict determinism is ultimately correct, the direction of thought was powerful. Events were to be explained by lawful causes, not by arbitrary supernatural decisions.

This mattered because divine will had often been used to explain unpredictability.

If something unexpected happened, God could be invoked. If something was too complex to understand, God could be placed there. If the future was unknown, divine mystery could be used to cover ignorance.

Laplace pushed the opposite habit.

Do not rush to God because the calculation is difficult.

Do not call something divine because it is complex.

Do not place supernatural agency where patient mathematics may one day reach.

This is scientific humility in a strong form.

It does not say, “I know everything.”

It says, “Do not pretend God explains what we have not yet understood.”

The Difference Between Mystery and Explanation

A mystery can be honest.

An explanation must do work.

This distinction is essential.

There are still mysteries in science. There are still questions about the origin of the universe, the nature of consciousness, the foundations of physical law, dark matter, dark energy, and the deepest structure of reality. A thoughtful person should not deny this.

But mystery alone is not an explanation.

To say, “We do not yet understand this fully,” is honest.

To say, “Therefore God did it,” goes beyond the evidence.

That is the problem.

God often enters the discussion at the point where human knowledge pauses. But the history of science teaches us caution. Many things once placed in divine territory later received natural explanation. Thunder, disease, eclipses, planetary motion, fossils, mental illness, and biological adaptation were all once surrounded by supernatural thinking.

The question is not whether mystery exists.

The question is whether mystery justifies God.

Laplace's answer was clear in his own field. Celestial mechanics did not need God. The mystery that remained was not solved by inserting a divine name into the equation.

A name is not a mechanism.

A sacred answer is not automatically a true answer.

An unknown cause is not evidence for a supernatural cause.

This principle will matter throughout the rest of the series.

The Shrinking Role of God

Laplace marks an important stage in the shrinking role of God.

In ancient religion, God or gods could explain nearly everything. Rain, thunder, fertility, plague, war, kingship, dreams, madness, death, and the stars could all be placed under divine power.

With Copernicus, Earth moved from the center.

With Galileo, the heavens lost their perfection.

With Kepler, planetary motion became mathematical.

With Newton, Earth and heaven came under one law.

With Laplace, God became unnecessary as a working scientific hypothesis for celestial mechanics.

This is not the end of religion, but it is a major change in religion's explanatory status.

The more science explained, the more God had to move.

God moved from the sky to the background of the sky.

From the orbit to the law behind the orbit.

From direct action to abstract source.

From explanation to interpretation.

This movement is often presented as sophistication. Sometimes it is. A mature religious thinker may no longer imagine God pushing planets or throwing lightning.

But we should also see what has happened.

Science forced the retreat.

Religion adapted because the older claims became harder to defend.

The God who once explained specific events became the God behind general existence. That may still be meaningful to believers, but it is less useful as an explanation of nature.

The Difference Between Belief and Necessity

This chapter is not arguing that no one can believe in God after Laplace.

Many people did. Many still do.

The question is different.

Is God necessary?

That word matters.

A person may believe that God created gravity. But gravity can be studied without that belief. A person may believe that God sustains the universe. But physics can work without that assumption. A person may believe that God is the reason there is order. But astronomy can calculate the order without invoking God.

The belief may remain.

The necessity disappears.

This distinction is central to the entire series.

A belief can survive in emotion, culture, theology, and philosophy after it has lost its role in scientific explanation. People may continue to pray, worship, sing, hope, fear, and believe. But the natural world no longer needs that belief in order to be explained.

That is why Laplace's answer is so powerful.

He did not say, "I have disproved God."

He said, in effect, "I do not need God for this explanation."

That is a more careful and more devastating statement.

Why "God Did It" Stops Inquiry

There is another reason Laplace matters.

The God hypothesis can stop investigation too early.

If we explain planetary motion by saying God moves the planets, there is little reason to develop celestial mechanics. If we explain disease by saying God punishes sin, there is little reason to search for microbes. If we explain lightning by saying God is angry, there is little reason to study electricity. If we explain species by saying God designed them separately, there is little reason to investigate evolution.

"God did it" often satisfies the mind before the mind has done the hard work.

This is dangerous.

It gives the appearance of explanation without the discipline of discovery.

Science begins when human beings resist that temptation. It begins when we say: perhaps there is a natural cause we do not yet understand. Let us look. Let us measure. Let us test. Let us calculate. Let us compare. Let us wait before declaring mystery solved.

This is why the sentence "I do not know" is often more honest than "God did it."

"I do not know" opens the road.

"God did it" often closes it.

Laplace stands for the open road.

Natural Explanation Proves Itself

Laplace's work belongs to a larger truth: natural explanation earns trust by performance.

Celestial mechanics did not merely sound elegant. It allowed calculation. It brought order to complexity. It helped explain planetary motion without invoking divine adjustment. It showed that the heavens could be studied with the same seriousness and precision as any other part of nature.

This practical success matters.

A religious explanation may inspire awe, but it does not calculate the perturbations of planets.

A theological statement may give meaning, but it does not predict celestial motion.

A sacred text may comfort a believer, but it does not improve an astronomical model.

Mathematics did.

Physics did.

Observation did.

Natural law did.

This is why God became unnecessary in this field. Not because human beings suddenly lost the capacity for wonder, but because the natural explanation worked better.

It explained more.

It predicted more.

It needed fewer unsupported assumptions.

It allowed correction.

Science did not win by shouting louder than religion.

It won by explaining the heavens more accurately.

The Religious Reply Grows More Abstract

As science advanced, many religious replies became more abstract.

Earlier religion could say: God sends the storm. God punishes through disease. God moves the heavens. God creates each species. God causes victory in war. God blesses the harvest.

After science, these claims became harder to maintain in their direct form.

So God moved upward, backward, and outward.

God became the creator of laws, the ground of existence, the source of being, the reason there is something rather than nothing, the explanation for why mathematics works, or the foundation of order.

Some of these are serious philosophical ideas. They deserve careful discussion, especially in later books. This book should not dismiss them casually.

But we should notice the pattern.

The more science explains the specific workings of nature, the more God becomes distant from those workings.

That distance matters.

A distant philosophical God may survive scientific explanation more easily than a God who directly sends lightning, disease, or planetary motion. But such a God also explains less of the everyday world that religion once claimed to explain.

This is why the history of science is also the history of theological retreat.

Not always the retreat of belief.

But the retreat of God as explanation.

What Laplace Teaches Us

Laplace teaches us a disciplined lesson:

Do not add unnecessary explanations.

If a natural explanation explains the evidence, predicts the event, and fits the facts, then adding God does not strengthen it. It may satisfy religious feeling, but it does not improve the science.

This principle is not anti-religious in a crude sense. It is simply intellectually disciplined.

A good explanation should not multiply causes without need. It should not add invisible agents where lawful causes already explain the event. It should not use mystery as a shortcut.

In daily life, we already understand this.

If a window breaks during a storm, and we find that a tree branch struck it, we do not need to add that a ghost pushed the branch.

If a patient develops pneumonia, and testing shows a bacterial infection, we do not need to add that a demon entered the lungs.

If an eclipse occurs exactly when celestial mechanics predicts it, we do not need to add that God sent it as a warning.

The natural explanation is enough.

This does not mean no other philosophical questions can be asked. But it does mean that the specific event has been explained.

God becomes an extra layer.

And an extra layer is not the same as necessity.

The Religious Reply

A thoughtful believer may say:

"Laplace may not have needed God in his equations, but equations do not explain why the universe exists, why there are laws, why mathematics applies to nature, or why there is order at all. Science explains mechanisms, but God explains the deeper foundation."

This is the strongest reply.

It correctly points out that celestial mechanics does not answer every philosophical question. Physics may describe how bodies move, but it does not automatically settle why there is a universe or why there are laws of nature. These deeper questions remain meaningful and will be discussed more fully in later books in this series.

But this reply also confirms the main argument of this chapter.

God is no longer needed to explain the motion of the planets.

God has been moved to a different question.

That move may preserve a kind of belief, but it gives up the older explanatory role. The God who once moved the heavens is replaced by mathematics and physics. The believer now argues for God behind existence itself, not as the direct explanation of celestial motion.

That is a retreat.

A sophisticated retreat perhaps, but still a retreat.

Evidence-Based Response

The evidence-based response is simple.

When a natural explanation works, we should not add a supernatural one without need.

Celestial mechanics explained planetary motion better than divine action. It gave mathematics, prediction, and correction. It allowed scientists to calculate and understand the heavens in ways religious explanation never could.

God did not improve the model.

God did not make the calculation more accurate.

God did not produce a clearer prediction.

God did not explain the mechanism.

Therefore, for that scientific purpose, God was unnecessary.

This is not a total disproof of every possible God. It is something more historically grounded: a demonstration that one more area of reality no longer required divine explanation.

Laplace's importance lies there.

He gave language to what science had increasingly shown.

God was not needed as a scientific hypothesis.

Chapter Summary

Laplace represents a major moment in the history of scientific explanation.

After Newton had brought Earth and heaven under one law, Laplace helped develop celestial mechanics in a way that made God unnecessary as part of the working explanation. The famous phrase, "I had no need of that hypothesis," captures the deeper point: if natural law explains the motion of the heavens, God does not need to be added.

This does not prove that no God can exist. It does not answer every philosophical question about existence, law, or order. But it does show that God was no longer needed to explain planetary motion.

God moved from direct explanation to background belief.

That movement is one of the central patterns of this book.

Key Points

Laplace sharpened the idea that God was unnecessary as a scientific hypothesis.

A useful scientific hypothesis must explain, predict, clarify, or guide investigation.

"God did it" often names a supposed cause without explaining the mechanism.

Celestial mechanics explained planetary motion without divine action.

God may still be claimed as the source of laws, but that is a different and more abstract claim.

Laplace shows the difference between personal belief and scientific necessity.

Transition to the Next Chapter

By the time of Laplace, the heavens had become increasingly lawful. Earth had moved from the center. The sky had lost its perfection. Planetary motion had become mathematical. Gravity had united Earth and heaven. Celestial mechanics had made God unnecessary as a working hypothesis.

But science was not finished.

The next great transformation would come closer to home.

If the heavens could be explained naturally, what about the human body itself?

For centuries, the body had been surrounded by mystery, sacred ideas, ancient authority, and mistaken theories. Slowly, anatomy and physiology would open the body to natural explanation.

In the next chapter, we will see how the body became natural.

Sources

Pierre-Simon Laplace, *Celestial Mechanics*

Pierre-Simon Laplace, *A Philosophical Essay on Probabilities*

Roger Hahn, *Pierre Simon Laplace, 1749–1827: A Determined Scientist*

Charles Coulston Gillispie, *Pierre-Simon Laplace, 1749–1827: A Life in Exact Science*

Isaac Newton, *Mathematical Principles of Natural Philosophy*

I. Bernard Cohen, *The Newtonian Revolution*

Alexandre Koyré, *From the Closed World to the Infinite Universe*

Stanford Encyclopedia of Philosophy, "Determinism: Causal Determinism"

Stanford Encyclopedia of Philosophy, "Newton's Philosophy"

John Hedley Brooke, *Science and Religion: Some Historical Perspectives*

Peter Harrison, *The Territories of Science and Religion*

Carl Sagan, *Cosmos*

Jacob Bronowski, *The Ascent of Man*

Chapter 8: When the Body Stopped Being Sacred Mystery

The Mystery Under the Skin

The human body is the most familiar mystery in the world.

We live inside it every moment. We feel hunger, pain, breath, heartbeat, sleep, weakness, desire, fear, and fatigue. We watch children grow. We see wounds bleed. We see fever burn the skin. We see pregnancy enlarge the abdomen. We see aging bend the back and slow the step. Finally, we see death still the body completely.

The body is near us.

Yet, for most of history, it was hidden from us.

People could see skin, eyes, hair, hands, sweat, tears, blood, wounds, swelling, and deformity. But they could not clearly see the inner world: organs, vessels, nerves, muscles, bones, lungs, heart, liver, kidneys, brain, and reproductive system. They could see symptoms, but not structure. They could see suffering, but not mechanism.

Because the inside of the body was hidden, imagination filled the space.

The body became a place of spirits, humors, vital forces, divine breath, moral purity, impurity, possession, punishment, and mystery. Illness was explained through religion, magic, imbalance, curse, sin, fate, or invisible spiritual forces. Death was explained as the departure of soul or spirit. Madness could be called possession. Seizures could be feared as supernatural. Childbirth was wrapped in danger, ritual, prayer, and superstition.

This was understandable.

Even today, after centuries of anatomy, physiology, pathology, microbiology, imaging, surgery, genetics, and intensive care, the body remains astonishingly complex. No honest physician thinks the body is simple.

But the great transformation began when human beings did something both simple and difficult.

They looked inside.

They opened the body.

They studied its structure.

They compared what they saw with what they had been taught.

They allowed the body itself to correct ancient authority.

That was a major step in making God unnecessary as an explanation.

Not unnecessary for grief. Not unnecessary for funeral rituals. Not unnecessary for comfort. But unnecessary as an explanation for how the body works.

The body slowly moved from sacred mystery to natural structure.

The Body Before Anatomy

Before modern anatomy, medical knowledge was a mixture of observation, tradition, philosophy, religious belief, and speculation.

Ancient physicians did observe patients carefully. They described symptoms. They identified patterns. They used herbs, diet, splints, cautery, surgery, and practical treatments. Some ancient medical traditions were thoughtful and impressive for their time.

But without systematic dissection and physiology, much remained hidden.

People did not know how blood circulated. They did not understand oxygen. They did not know about cells. They did not understand hormones. They did not know how kidneys filtered blood. They did not understand the nervous system in the modern sense. They did not know how the brain produced speech, movement, memory, emotion, and personality. They did not know about bacteria and viruses, though that subject belongs more fully to a later chapter.

So they explained the body with the concepts available to them.

In some traditions, the body was governed by humors, elements, energies, spirits, or vital forces. Health meant balance. Disease meant imbalance, impurity, invasion, divine displeasure, or disorder. Some ideas had practical value. Others were deeply mistaken.

The problem was not only ignorance.

It was authority.

Once a great ancient teacher said something, later generations often repeated it. Books were copied. Ideas were preserved. Students were trained to respect old masters. Religious and cultural restrictions sometimes limited the direct study of human bodies. Dissection could be seen as disrespectful, dangerous, impure, or spiritually troubling.

The body, in many minds, was not simply a biological organism.

It was connected to the soul. It was connected to purity. It was connected to death. It was connected to God.

That made investigation difficult.

Yet medicine could not truly advance until the body was allowed to speak for itself.

The Dead Body as Teacher

A dead body can disturb us.

It once held a person. It once had warmth, breath, voice, appetite, fear, memory, pain, hope, and family. To cut into it can feel emotionally difficult even today. In earlier religious cultures, the fear was often stronger.

Was the body needed for resurrection?

Would dissection dishonor the dead?

Would cutting the body disturb the soul?

Was the corpse impure?

Was touching the dead spiritually dangerous?

Was the body sacred property under divine ownership?

These questions were not trivial to people who believed deeply in soul, resurrection, purity, and divine judgment.

But the living were suffering.

Women died in childbirth. Wounds became fatal. Internal disease remained mysterious. Surgical practice was limited. Mistaken anatomy led to mistaken treatment. Physicians often repeated old ideas because they had not seen the body directly enough.

Medicine faced a moral tension.

Should the dead body remain untouched because of sacred fear, or should it teach the living how to suffer less?

Modern medicine chose learning.

That choice was not always clean or morally pure. The history of anatomy includes abuses, grave robbing, exploitation of the poor, and unethical practices. We should not romanticize it. Human progress often comes mixed with human failure.

But the central intellectual movement remains important.

The body became available to evidence.

Once that happened, mystical and sacred ideas of bodily function began to lose authority.

The dead body became a difficult teacher.

And the living benefited.

Galen and the Power of Ancient Authority

For more than a thousand years, the writings of Galen strongly influenced medicine in Europe and the Islamic world.

Galen was a brilliant physician of antiquity. He made many observations and contributed significantly to medical thought. He studied anatomy, often through animal dissection rather than human dissection. Because he could not freely dissect human bodies in the way later anatomists would, some of his conclusions were wrong when applied to humans.

Yet his authority became enormous.

For centuries, students learned medicine through texts heavily shaped by Galen. His ideas about anatomy, physiology, humors, and bodily function were treated with great respect. In many settings, what Galen had written became more powerful than what the human body itself might reveal.

This is a familiar pattern.

Authority can preserve knowledge, but it can also preserve error.

A revered teacher may be right about many things and wrong about others. The danger begins when respect becomes obedience, and obedience becomes intellectual paralysis.

Science requires a different attitude.

Honor the past, but test it.

Respect the teacher, but look at reality.

Read the book, but examine the body.

Do not let authority silence evidence.

This was the spirit that would eventually transform anatomy.

The body had to be seen not as a sacred object protected from investigation, nor as a philosophical idea described by ancient authority, but as a natural structure that could be studied directly.

Vesalius and the Courage to Look

Andreas Vesalius was one of the great turning points in the history of anatomy.

In 1543, the same year Copernicus published *On the Revolutions of the Heavenly Spheres*, Vesalius published *De Humani Corporis Fabrica*, usually known in English as *On the Fabric of the Human Body*.

The timing is striking.

Copernicus was moving the Earth from the center of the cosmos.

Vesalius was opening the human body to direct observation.

One changed the sky.

The other changed the body.

Vesalius insisted on looking for himself. He performed dissections. He studied human anatomy directly. He compared what he saw with what had been taught. When the body contradicted Galen, Vesalius increasingly trusted the body.

This was a profound act of intellectual courage.

The lesson was not merely anatomical. It was philosophical.

The human body itself became the authority.

If the ancient text said one thing and the dissected body showed another, the body had to win.

That is the scientific attitude in its clearest form.

Reality over tradition.

Evidence over authority.

Observation over inherited certainty.

Vesalius did not understand everything. No scientist does. But his work helped free anatomy from blind dependence on ancient texts. His detailed illustrations gave students and physicians a new way to see the body. Organs, muscles, bones, vessels, and structures became visible in a more systematic way.

The body was no longer only a vessel of mystery.

It was a structure.

And structure can be studied.

Harvey and the Motion of Blood

If Vesalius helped reveal the structure of the body, William Harvey helped reveal one of its great functions.

In 1628, Harvey published *On the Motion of the Heart and Blood.* He argued that blood circulates through the body, pumped by the heart, moving through arteries and veins in a closed system.

This seems obvious to us now.

But it was revolutionary.

For a long time, older ideas about blood and bodily fluids had dominated medical thought. Blood was not clearly understood as circulating continuously. The heart's function, the movement of blood, and the relationship between arteries and veins were misunderstood.

Harvey used observation, measurement, and experiment. He studied the heart. He examined valves in veins. He estimated the amount of blood moved by the heart and saw that the body could not simply be producing and consuming blood in the older way imagined. The numbers did not fit.

This is a beautiful example of scientific reasoning.

Observation mattered.

Anatomy mattered.

Experiment mattered.

Measurement mattered.

Logic mattered.

The body was not explained by sacred symbolism. It was explained by structure and function.

The heart was not merely the seat of courage, love, spirit, or emotion.

It was a muscular pump.

That sentence may sound cold, but it is medically powerful.

To understand the heart as a pump does not make love less real. It does not make courage meaningless. It does not reduce human life to machinery in a crude sense. But it does identify the physical function of the organ.

And once function is understood, disease can be understood more clearly.

If the heart pumps blood, then heart failure, valve disease, blocked arteries, shock, bleeding, and circulation can be studied in a new way. The body becomes less mysterious and more treatable.

The Heart Without Myth

The heart has always carried symbolic meaning.

We speak of a kind heart, a broken heart, a brave heart, a heavy heart. Across cultures, the heart has been connected with love, courage, sincerity, emotion, and the inner self.

These meanings remain powerful in poetry and ordinary language.

But medicine had to separate symbol from function.

The heart as symbol belongs to literature, emotion, and culture.

The heart as organ belongs to anatomy and physiology.

Confusing the two can be dangerous.

If the heart is treated mainly as a mystical center of spirit, we may miss its physical reality. If chest pain is interpreted as emotional weakness or divine testing, we may miss a heart attack. If shortness of breath and swelling are treated only with prayer, we may miss heart failure. If sudden collapse is seen as fate, we may miss arrhythmia, bleeding, pulmonary embolism, or shock.

Modern medicine works because it respects the physical body.

It does not deny that human beings have emotions, meanings, and stories. But it does not allow symbolic meaning to replace biological explanation.

This is the same pattern seen throughout this book.

The sky can inspire poetry, but astronomy explains planetary motion.

The Sun can symbolize life, but physics explains its energy.

The heart can symbolize love, but physiology explains circulation.

Disease can raise spiritual questions, but biology explains infection.

Science does not forbid meaning.

It prevents meaning from pretending to be mechanism.

The Body as Structure and Function

Anatomy revealed structure.

Physiology revealed function.

Together, they changed medicine.

The body was no longer a vague container of spirit, humors, and mysterious forces. It became a living system with parts that could be studied. Bones supported. Muscles moved. The heart pumped. Blood circulated. Lungs exchanged gases. Kidneys filtered. Nerves conducted signals. The brain coordinated movement, sensation, speech, memory, and behavior.

This was not discovered all at once. It took centuries. Many mistakes were made. Many details came later. But the direction was clear.

The body became natural.

That word is important.

Natural does not mean simple. The body is not simple. It is more complex than ancient people could have imagined.

Natural does not mean meaningless. A natural body can still be precious, personal, and worthy of deep respect.

Natural does not mean fully understood. Medicine still has limits.

Natural means that the body works through physical, chemical, biological, and physiological processes, not through invisible spirits or divine manipulation in ordinary function.

This changed the human view of illness.

If the body is natural, then disease is natural too. It may be tragic. It may be painful. It may be emotionally devastating. But it is not automatically a curse, punishment, possession, or spiritual stain.

That shift is morally important.

It allows compassion to replace blame.

Disease Without Moral Blame

One of the cruelest effects of supernatural thinking is that it can blame the sick.

If illness is punishment, the patient becomes morally suspect. If disease is impurity, the patient becomes spiritually contaminated. If seizures are possession, the patient becomes frightening. If infertility is divine disfavor, the woman may be shamed. If mental illness is demonic, the suffering person may be abused instead of treated.

This is not only a problem of the past. Even today, some people still attach moral or spiritual blame to illness.

As a physician, I find this deeply troubling.

Sick people need care. They do not need accusation. They need diagnosis, treatment, comfort, honesty, and human dignity. They do not need to be told that their suffering is proof of sin, weak faith, bad karma, curse, possession, or divine punishment.

A natural understanding of the body helps protect patients from this cruelty.

If diabetes is a disorder of metabolism, insulin, body weight, genetics, diet, and physiology, then it is not a divine curse. If epilepsy involves abnormal electrical activity in the brain, it is not possession. If depression involves brain function, psychology, life circumstances, genetics, and biology, it is not moral failure. If infection comes from microbes, it is not spiritual impurity.

Natural explanation can be humane.

It removes false guilt.

It allows the patient to be treated as a person, not as a symbol.

This is one of the great moral achievements of science-based medicine.

Natural Explanation Proves Itself

A natural understanding of the body did not remain a theory.

It worked.

Anatomy guided surgery. Physiology explained organ function. Pathology connected symptoms with tissue changes. Blood circulation helped physicians understand shock, bleeding, heart failure, and vascular disease. Knowledge of nerves helped explain paralysis, pain, sensation, and movement. Knowledge of lungs helped explain breathing failure. Knowledge of kidneys helped explain fluid balance, toxins, and electrolyte disorders.

This practical success is important.

A prayer may comfort a patient, and comfort has its own human value.

But prayer does not tell a surgeon where an artery lies.

Prayer does not show how blood flows through a damaged valve.

Prayer does not explain why a blocked coronary artery can kill heart muscle.

Prayer does not reveal why kidney failure causes potassium to rise.

Anatomy and physiology do that.

Modern medicine became powerful because it studied the body as nature. It did not treat the body as a fog of sacred mystery. It did not stop at inherited authority. It looked, measured, compared, corrected, and learned.

This is why the body becoming natural was such an important step in the retreat of divine explanation.

The body did not become less wonderful.

It became more understandable.

And because it became more understandable, it became more treatable.

The Brain and the Self

Even in the early history of anatomy, the brain presented a special challenge.

The brain was not just another organ. It was connected to sensation, movement, memory, speech, thought, emotion, and identity. The more medicine studied the nervous system, the harder it became to treat the mind as something entirely separate from the body.

This topic will be explored more fully later in this book and in later volumes. But it belongs here in a limited way.

The study of the body began to narrow the space for the supernatural soul.

Brain injury could change speech.

Brain disease could change personality.

Stroke could paralyze one side of the body.

Epilepsy could produce seizures and altered awareness.

Dementia could damage memory and identity.

Tumors, trauma, infection, and degeneration could change thought and behavior.

These facts are difficult for a simple soul-body division.

If the self were fully independent of the body, why would injury to the brain change the self so profoundly? Why would damage to one region disturb speech and another region affect movement? Why would chemicals, oxygen, blood flow, infection, or trauma alter consciousness?

Again, this does not settle every philosophical question about consciousness. We should not overclaim. But it does show the direction of evidence.

The mind is deeply dependent on the body.

And especially on the brain.

This makes the old idea of a fully separate, immaterial soul less necessary and less convincing as an explanation of ordinary mental life.

Medicine Learns to Trust the Body

Modern medicine rests on a simple principle: the body must be studied as it is.

Not as we imagine it.

Not as tradition describes it.

Not as scripture suggests.

Not as ancient authority claimed.

Not as superstition fears it to be.

As it is.

This is why medical training requires anatomy, physiology, pathology, microbiology, pharmacology, imaging, clinical examination, laboratory testing, and evidence-based treatment. The physician must learn from the body itself.

The patient's story matters. The physical examination matters. The blood test matters. The X-ray matters. The biopsy matters. The heart tracing matters. The culture result matters. The response to treatment matters.

Medicine becomes dangerous when it ignores reality.

If a doctor treats what he wishes were true instead of what is true, the patient suffers.

This is why the scientific attitude is not merely academic. It is ethical.

A physician has a moral duty to follow evidence because a human life may depend on it.

That is why medicine is such a powerful example in this series. When supernatural explanations fail, people suffer. When natural explanations succeed, people may live.

The body becomes natural, and treatment becomes possible.

The Religious Reply

A thoughtful believer may say:

"Anatomy and physiology explain how the body works, but they do not remove God. God may have created the body. God may have designed its organs. God may work through natural processes. The body's complexity may even point toward divine wisdom."

This reply is understandable.

The human body is astonishing. Its organs, tissues, cells, blood vessels, nerves, immune responses, hormones, and repair mechanisms can inspire wonder. A believer may look at the body and feel gratitude to God.

But the question in this book is more precise.

Is God needed to explain how the body works?

In medicine, the answer is no.

We do not explain circulation by invoking divine breath. We explain it through the heart, blood vessels, pressure, valves, and physiology. We do not explain seizures by invoking demons. We explain them through abnormal electrical activity in the brain. We do not explain kidney failure by invoking impurity. We explain it through filtration, blood flow, nephron function, inflammation, diabetes, hypertension, toxins, obstruction, and other natural causes.

God may remain as personal belief, but God does not do the explanatory work.

Anatomy and physiology do.

Evidence-Based Response

The evidence-based response is clear.

The body became understandable when it was studied directly.

Dissection corrected ancient errors. Anatomy revealed structure. Physiology revealed function. Experiment and measurement replaced speculation. The heart became a pump. Blood became a circulating fluid. Organs became physical systems. Disease became increasingly tied to natural causes.

This did not remove all mystery from the body. The body remains complex and sometimes difficult to understand. Medicine still has limits. Physicians still face uncertainty. Patients still suffer. Death still comes.

But the direction of progress is unmistakable.

Natural explanation has done what supernatural explanation did not do. It has allowed diagnosis, surgery, anesthesia, antisepsis, imaging, transfusion, organ support, intensive care, and modern treatment. It has helped reduce suffering and save lives.

A prayer may comfort a patient.

But anatomy tells the surgeon where to cut.

Physiology tells the physician what is failing.

Pathology tells us what disease is present.

Microbiology tells us what organism may be responsible.

Pharmacology tells us what medicine may help.

The body became natural.

And when the body became natural, God became less necessary as an explanation of bodily life.

Chapter Summary

For centuries, the human body was surrounded by mystery, sacred fear, ancient authority, and supernatural explanation. Disease, death, madness, childbirth, and bodily suffering were often interpreted through spirits, humors, divine punishment, impurity, or fate.

The rise of anatomy and physiology changed this. Vesalius helped correct ancient anatomical errors through direct study of the human body. Harvey showed that blood circulates and that the heart functions as a pump. Over time, the body became understood as a natural system of organs, tissues, vessels, nerves, and functions.

This transformation did not make the body less wonderful. It made the body more understandable.

It also weakened supernatural explanation. The body no longer needed to be explained through spirits, curses, divine punishment, or mystical forces. It could be studied as nature.

That made medicine more honest, more effective, and often more compassionate.

Key Points

Before modern anatomy, the body was often explained through spirits, humors, vital forces, divine punishment, impurity, or mystery.

Ancient authorities such as Galen preserved important knowledge but also preserved errors.

Vesalius helped shift anatomy from textual authority to direct observation of the human body.

Harvey showed that blood circulates and that the heart works as a pump.

Anatomy and physiology moved the body from sacred mystery to natural explanation.

Natural explanations of disease reduce false blame and allow more compassionate care.

The body remains wonderful, but it does not require God as a working explanation.

Transition to the Next Chapter

The human body was not the only mystery that slowly became natural.

Matter itself was also mysterious.

For centuries, people wondered what things were made of, why substances burned, why metals changed, why gases behaved strangely, and whether matter could be transformed through hidden powers. Alchemy, magic, philosophy, and early chemistry all mixed together in the human search to understand substance.

The next great step came when matter, like the body, was brought under experiment and measurement.

In the next chapter, we will see how chemistry helped replace magic with natural law.

Sources

Andreas Vesalius, *De Humani Corporis Fabrica*

William Harvey, *On the Motion of the Heart and Blood*

Galen, *On the Natural Faculties*

Charles Singer, *A Short History of Anatomy and Physiology from the Greeks to Harvey*

Andrew Cunningham, *The Anatomical Renaissance*

Vivian Nutton, *Ancient Medicine*

Roy Porter, *The Greatest Benefit to Mankind*

Sherwin B. Nuland, *Doctors: The Biography of Medicine*

Jacalyn Duffin, *History of Medicine: A Scandalously Short Introduction*

Encyclopedia Britannica, "Andreas Vesalius"

Encyclopedia Britannica, "William Harvey"

Stanford Encyclopedia of Philosophy, "Medicine, Philosophy of"

John Hedley Brooke, *Science and Religion: Some Historical Perspectives*

Chapter 9: Matter Without Magic

The Fire in the Furnace

Imagine an old workshop at night.

A fire burns inside a furnace. Metal glows red, then orange, then almost white. Smoke rises. Liquids boil. Powders change color. A hard mineral softens. A dull substance becomes bright. A clear liquid turns cloudy. Something solid becomes vapor. Something ordinary suddenly behaves strangely.

To a mind without modern chemistry, such changes could feel almost magical.

Matter seemed to hide secrets.

Wood became ash. Iron rusted. Grapes became wine. Bread rose. Milk soured. Metals melted. Medicines healed. Poisons killed. Water disappeared into steam. Flesh decayed. Gold remained bright while other metals corroded and darkened.

What was happening inside these transformations?

Before chemistry became a science, people could see the change, but not the mechanism. They could use fire, minerals, herbs, dyes, metals, and medicines, but they did not understand atoms, molecules, oxygen, gases, acids, bases, reactions, enzymes, fermentation, or conservation of matter.

So matter became surrounded by mystery.

It was explained through elements, essences, spirits, qualities, humors, occult forces, divine signatures, and hidden sympathies. Alchemy mixed observation with magic, practical craft with mystical hope, experiment with symbolism. Some alchemists were careful observers. Some developed useful techniques. But their explanations were often wrapped in secrecy, metaphor, and spiritual meaning.

Matter was not yet fully natural.

It was a stage for mystery.

Chemistry slowly changed that. It did not happen in one sudden moment. It took centuries of measurement, experiment, better instruments, clearer language, and intellectual courage. But eventually, matter moved from magic to science.

The transformation of chemistry is another major chapter in the retreat of divine and supernatural explanation.

When Change Looked Magical

Transformation has always fascinated human beings.

A seed becomes a tree. Grapes become wine. Dough becomes bread. Ore becomes metal. Fire turns wood into smoke, heat, flame, and ash. A clear liquid can become colored. A medicine can quiet pain. A poison can stop breathing.

These transformations feel almost magical because the cause is hidden.

The eye sees the result, but not the underlying process.

Before atoms, molecules, oxygen, carbon, acids, bases, reactions, enzymes, and microbes were understood, material change seemed mysterious. People could learn by experience that some processes worked, but they could not clearly explain why.

This is fertile ground for supernatural thinking.

When a process works but is not understood, people often surround it with ritual. They give it symbolic meaning. They hide it behind secret language. They connect it to cosmic forces, planets, gods, purity, corruption, or spiritual power.

Alchemy grew in this kind of world.

It was not merely silly magic, as it is sometimes portrayed. It was a complicated historical mixture. It included practical laboratory work, metallurgy, medicine, dye-making, distillation, sublimation, crystallization, and metal refining. But it also included mystical dreams: the purification of matter, the transmutation of metals, the philosopher's stone, and the search for an elixir of life.

The alchemist's furnace was both laboratory and temple.

Matter was studied, but also spiritualized.

Science had to separate these things.

It had to ask: What is actually happening to the substance?

Not what does it symbolize?

Not what hidden spirit controls it?

Not what sacred meaning is attached to it?

Not what ancient authority claimed?

What happens when we measure, heat, weigh, combine, separate, and repeat?

That question was the beginning of chemistry.

The Four Elements and the Comfort of Simplicity

For a long time, many thinkers explained matter through the idea of a few basic elements.

In Greek thought, the familiar four were earth, water, air, and fire. Other cultures had their own elemental systems. These ideas were appealing because they made the material world feel understandable. The variety of nature could be reduced to a small number of basic principles.

Earth was solid.

Water was liquid.

Air was invisible but felt.

Fire transformed.

This scheme had beauty. It matched ordinary experience. It gave people a way to classify the world. It was simple, memorable, and philosophically satisfying.

But it was not modern chemistry.

The four elements were not elements in the scientific sense. They were broad qualities of experience, not measurable chemical substances. They could not explain why oxygen supports combustion, why hydrogen and oxygen form water, why acids react with bases, why metals oxidize, or why carbon chemistry produces such complex life.

The old elemental theory survived for a long time because it felt elegant.

This is another lesson in science.

A beautiful idea can be wrong.

Human beings often prefer simple pictures. We like systems that feel complete. We like explanations that fit neatly into memory and tradition. But nature is not required to be simple in the way we prefer.

Chemistry forced the human mind to leave behind poetic categories and enter the harder world of measurement.

It was not enough to say that fire was an element. One had to ask what burning actually is.

It was not enough to say that air was one thing. One had to discover that air is a mixture.

It was not enough to say that matter had hidden qualities. One had to identify substances, reactions, and conservation.

Chemistry became science when matter was allowed to correct philosophy.

Alchemy: Half Dream, Half Laboratory

Alchemy deserves a fair treatment.

It is easy to mock it because many of its goals now seem impossible or misguided. Turning lead into gold, finding the philosopher's stone, or discovering an elixir of immortality are not scientific goals in the modern sense. They belong more to mystical hope than reliable chemistry.

But alchemy also preserved practical knowledge.

Alchemists worked with furnaces, vessels, minerals, acids, distillation, sublimation, crystallization, and metal refining. They learned techniques that later chemists could use. They observed changes in matter. They developed laboratory habits, even if their theories were often wrong.

This is how human knowledge often grows.

The path from superstition to science is not always clean. Useful observation can live beside false theory. Practical skill can coexist with magical thinking. A person can discover something real while explaining it badly.

Alchemy was like that.

It showed that matter could be manipulated, but it did not yet provide the clear natural explanations that modern chemistry required.

The shift from alchemy to chemistry required a new discipline: open reporting, careful measurement, repeatable experiment, clearer definitions, and rejection of secret mystical language.

Secrecy was natural to alchemy.

Public correction was essential to chemistry.

That difference matters.

A secret tradition protects authority.

A scientific method exposes claims to testing.

Alchemy often asked how matter could be spiritually perfected. Chemistry asked what matter is and how it reacts.

The second question proved more powerful.

Boyle and the New Chemical Mind

Robert Boyle was one of the important figures in the movement from alchemy toward chemistry.

In *The Sceptical Chymist*, published in 1661, Boyle challenged older ideas about elements and chemical explanation. He argued for a more experimental approach to matter. He was skeptical of traditional systems that explained substances through broad qualities or inherited categories without sufficient evidence.

Boyle's importance lies not only in specific discoveries, but in attitude.

He wanted chemistry to become more disciplined. He wanted claims to be tested. He wanted matter studied through experiment rather than accepted through ancient authority. He helped push chemistry toward a more modern scientific form.

Boyle was also personally religious. This is another example of the pattern we have seen before.

A religious person can help create a natural explanation.

Boyle did not need to be an atheist to weaken magical thinking about matter. His commitment to experiment made the difference. He helped move the study of substance away from mystical speculation and toward observable behavior.

This is one of the important subtleties of this book.

Science did not advance only because atheists attacked religion. Many believers helped build science. But the methods they used reduced the need for supernatural explanations.

Boyle's faith did not prevent him from insisting that matter should be studied by experiment.

That insistence helped make chemistry less magical.

Air Was Not One Thing

For ordinary human experience, air seems simple.

It is invisible. It surrounds us. We breathe it. Wind moves it. Smoke rises into it. Birds fly through it. Clouds form in it. Fire burns in it.

For a long time, air was treated as one of the basic elements. It seemed like a single substance.

Chemistry showed otherwise.

Air is not one thing. It is a mixture. It contains nitrogen, oxygen, carbon dioxide, water vapor, and other gases in smaller amounts. Different gases have different properties. Some support life and combustion. Some do not. Some are produced in breathing, burning, fermentation, and decay.

This discovery changed how people understood life and matter.

Breathing was no longer just the movement of a vague life-force. Combustion was no longer simply the release of fiery essence. Fermentation was no longer a mysterious transformation guided by hidden powers. These processes could be studied as interactions of substances.

The discovery of gases helped chemistry enter the invisible world.

This is important because supernatural explanations often survive where causes are invisible. If the cause cannot be seen, imagination can easily fill the space.

But science repeatedly made the invisible visible, or at least measurable.

The telescope extended vision outward.

The microscope extended vision inward.

Chemistry extended understanding into invisible gases and reactions.

Once invisible causes became measurable, supernatural explanations lost ground.

Lavoisier and the End of Phlogiston

One of the great turning points in chemistry came with Antoine Lavoisier.

Before Lavoisier, many chemists explained burning through a theory called phlogiston. According to this older idea, combustible materials contained a fire-like substance called phlogiston that was released during burning.

The theory was wrong, but it had seemed useful for a time.

Lavoisier helped replace it with a better explanation based on oxygen and careful measurement. He showed the importance of weighing substances before and after reactions. He helped establish the principle that matter is conserved in chemical reactions. In ordinary chemical change, matter does not simply vanish. It changes form and combination.

This was revolutionary.

Burning was no longer the escape of a mysterious fire principle. It was a reaction involving oxygen.

Rusting, burning, and respiration could be understood in more natural and connected ways.

The material world became less magical and more lawful.

Lavoisier's work also shows the importance of measurement. Without weighing, chemistry remained vulnerable to vague theory. With weighing, matter had to obey evidence.

The balance became a weapon against mystery.

If a theory says something is escaping during combustion, but careful measurement shows mass increasing because oxygen combines with the substance, the theory must change.

This is how science advances.

Not by reverence for old explanation, but by obedience to evidence.

The Conservation of Matter

The idea that matter is conserved was deeply important.

In ordinary life, things seem to disappear. Wood burns and becomes ash and smoke. Water boils and vanishes. Food is digested and transformed. A candle burns down. Metal corrodes. A body decays.

Before chemical understanding, these changes could seem like disappearance, destruction, purification, or transformation through hidden powers.

Chemistry showed that matter changes form, combines, separates, and rearranges. It does not simply vanish in ordinary reactions.

This gave human beings a more stable understanding of the physical world.

The universe became less like a stage of magical transformation and more like a system of lawful change.

This did not remove wonder. In fact, it deepened wonder.

It is astonishing that wood, flame, air, heat, ash, smoke, and gas are connected through chemical reactions. It is astonishing that breathing, burning, and metabolism are related at the level of matter and energy. It is astonishing that the same kinds of atoms can form blood, bone, water, air, stone, poison, medicine, and stars.

Science does not make matter dull.

It makes matter intelligible.

The older magical view saw transformation and imagined hidden spirits. Chemistry saw transformation and searched for reactions.

The second approach gave us modern medicine, materials, agriculture, sanitation, anesthesia, antiseptics, fertilizers, fuels, and countless tools that changed human life.

Chemistry and the Body

Chemistry also changed medicine.

Before modern chemistry, medicines were often based on tradition, trial, plant use, mineral remedies, magical associations, and religious or folk belief. Some treatments helped. Many did not. Some were harmful.

Chemistry allowed medicine to become more precise.

It helped identify active substances. It helped develop drugs. It helped explain oxygen, carbon dioxide, blood chemistry, metabolism, acidity, electrolytes, hormones, toxins, anesthesia, antiseptics, and antibiotics. It helped show that the body is not only anatomical, but chemical.

As a physician, I see this every day.

A patient's sodium level matters.

Oxygen level matters.

Blood sugar matters.

Kidney function matters.

Acid-base balance matters.

Drug dose matters.

Toxic levels matter.

These are not mystical signs. They are measurable realities.

When a patient is critically ill, we do not ask whether an evil spirit has disturbed the salt in the blood. We measure sodium. We do not ask whether a demon has reduced oxygen. We measure oxygen and examine lungs, heart, blood, and circulation. We do not treat diabetic ketoacidosis with ritual. We treat it by understanding insulin, glucose, fluids, electrolytes, and acid-base chemistry.

Chemistry made the body more understandable and medicine more effective.

This is another way science made God unnecessary.

Not by removing human compassion, but by giving compassion knowledge.

Natural Explanation Proves Itself

Chemistry did not remain inside laboratories and books.

It changed ordinary life.

It helped create medicines. It helped make anesthesia possible. It helped improve sanitation. It helped explain nutrition, metabolism, poisoning, fermentation, combustion, corrosion, agriculture, and materials. It helped us understand what happens when a drug enters the body, when oxygen reaches the blood, when acid builds up, when toxins accumulate, and when chemical reactions support life.

This practical success matters.

A magical explanation may sound mysterious and impressive, but it does not tell us the safe dose of a medicine.

A ritual may give emotional comfort, but it does not explain why a poison stops breathing.

A sacred belief may assign meaning to water, fire, oil, wine, or blood, but it does not tell us their chemical properties.

Chemistry does.

It tells us what substances are, how they react, what they can do, and how they may harm or heal.

This is why chemistry weakened supernatural explanation. It did not merely deny magic. It replaced magic with knowledge that worked.

A correct chemical explanation can prevent poisoning, preserve food, sterilize equipment, produce medicine, and save lives.

That is more than intellectual progress.

It is human progress.

From Sacred Substance to Chemical Substance

Many cultures attached sacred meaning to substances.

Blood, water, wine, oil, incense, salt, ash, bread, herbs, metals, stones, and fire could carry ritual significance. They could purify, bless, protect, heal, curse, or symbolize divine power.

These meanings may still matter culturally and emotionally. A ritual object can be meaningful to a community. A symbol can carry memory and identity. A sacred meal can unite people. Water can symbolize cleansing. Fire can symbolize transformation.

But chemistry asks a different question.

What is the substance made of?

What are its properties?

How does it react?

What dose is safe?

What dose is toxic?

What happens when it enters the body?

Can its effect be tested?

This distinction is vital.

A substance may be sacred in a ritual and still have ordinary chemical properties. A plant may be used in a religious ceremony and still work or fail because of its active compounds. A poison may be surrounded by myth, but it kills through biology and chemistry. A medicine may be blessed, but its effect depends on pharmacology.

Science does not prevent people from giving symbolic meaning to substances.

It prevents symbolic meaning from replacing material truth.

That is why chemistry matters in this book.

It shows that matter does not behave according to prayer, ritual status, or sacred story. It behaves according to its properties and reactions.

The Decline of Magical Thinking About Matter

Magical thinking about matter did not disappear all at once.

Even today, people are drawn to the idea that substances carry hidden spiritual energies, that crystals heal by mystical vibrations, that water can remember blessings, that detox rituals remove vague impurities, or that natural substances are automatically safe because nature is imagined as benevolent.

These beliefs survive because matter still feels mysterious to many people.

Chemistry is difficult. The invisible world of atoms and molecules is not emotionally easy to grasp. A symbolic explanation may feel warmer than a chemical one. A ritual may feel more meaningful than a laboratory result.

But feeling is not evidence.

A crystal may be beautiful, but beauty is not proof of healing power. A substance may be natural, but natural does not mean safe. A ritual may be comforting, but comfort is not chemistry. A traditional remedy may be worth studying, but tradition alone does not prove effectiveness.

Modern chemistry teaches caution.

Substances have real properties. Dose matters. Purity matters. Interaction matters. Mechanism matters. Evidence matters.

This attitude has saved lives.

It has also weakened supernatural claims about matter. The more we understand the chemical world, the less we need invisible essences, magical forces, divine signatures, or spiritual energies to explain material change.

Matter remains wondrous.

But it is not magic.

The Religious Reply

A thoughtful believer may say:

"Chemistry explains how matter changes, but it does not explain why matter exists or why it follows laws. God may still be the creator of matter and the author of chemical order. The fact that matter is lawful may even point to divine intelligence."

This is a serious reply.

It is true that chemistry does not answer every philosophical question. It does not explain why there is something rather than nothing. It does not settle the ultimate origin of natural law. It does not tell us the final meaning of existence.

But that is not the question of this chapter.

The question is whether God is needed to explain chemical change.

The answer is no.

We do not need God to explain combustion.

We do not need God to explain rust.

We do not need God to explain fermentation.

We do not need God to explain oxygen.

We do not need God to explain acids and bases.

We do not need God to explain drug action.

We do not need God to explain why matter combines, separates, reacts, and changes.

Chemistry does that work.

A believer may place God behind the existence of matter. But God is not needed in the chemical explanation itself.

Once again, God moves from direct explanation to background belief.

That is the retreat.

Evidence-Based Response

The evidence-based response is clear.

Matter became understandable when people stopped treating transformation as magic and began studying it through experiment and measurement.

Boyle helped challenge older chemical traditions and encouraged a more experimental approach. Lavoisier helped replace phlogiston with oxygen-based chemistry and emphasized careful measurement. The discovery of gases, conservation of matter, and chemical reactions transformed the human understanding of substance.

Chemistry did not answer every ultimate question, but it explained material change far better than alchemy, magic, ritual, or divine action.

It gave mechanisms.

It gave measurements.

It gave predictions.

It gave useful knowledge.

It gave medicine new power.

It gave industry and agriculture new tools.

It made invisible processes intelligible.

Matter no longer needed spirits, essences, or divine manipulation to explain its ordinary behavior.

It became natural.

Chapter Summary

Before chemistry, matter was surrounded by mystery. Fire, fermentation, rust, medicine, poison, metal, vapor, and transformation were often explained through elements, essences, hidden powers, alchemy, and magical thinking.

Alchemy mixed useful practical work with mystical ideas. It preserved techniques and observations, but it did not yet provide a reliable scientific explanation of matter.

Modern chemistry emerged through experiment, measurement, clearer language, and correction of older theories. Boyle helped move chemistry away from inherited categories and toward experiment. Lavoisier helped explain combustion through oxygen and emphasized conservation of matter.

Chemistry made matter lawful and intelligible. It also changed medicine by making the body understandable at a chemical level.

This did not remove wonder from matter. It removed magic from explanation.

Key Points

Before modern chemistry, matter was often explained through elements, essences, hidden powers, alchemy, and magical thinking.

Alchemy contributed techniques and observations, but its explanations were often mystical and unreliable.

Boyle helped push chemistry toward experiment and skepticism of inherited theories.

Lavoisier helped replace phlogiston with oxygen-based chemistry and emphasized measurement.

Chemistry showed that matter changes through natural reactions, not magical or divine forces.

Chemical knowledge transformed medicine, agriculture, industry, and daily life.

God may remain a belief about the origin of matter, but God is not needed to explain ordinary chemical change.

Transition to the Next Chapter

Science had now changed the sky, the heavens, the movement of planets, the human body, and the behavior of matter.

But another great mystery remained beneath human feet.

How old is the Earth?

For many religious traditions, Earth's history was tied to sacred stories, genealogies, floods, creation accounts, and divine timelines. But rocks carried a different record. Mountains, fossils, sediments, erosion, and volcanic layers told a story far older than human memory and far deeper than sacred chronology.

In the next chapter, we will meet Hutton, Lyell, and the discovery of deep time.

Sources

Robert Boyle, *The Sceptical Chymist*

Antoine Lavoisier, *Elements of Chemistry*

Antoine Lavoisier, *Memoir on Combustion*

William H. Brock, *The Norton History of Chemistry*

Bernadette Bensaude-Vincent and Isabelle Stengers, *A History of Chemistry*

Jan Golinski, *Science as Public Culture: Chemistry and Enlightenment in Britain*

Lawrence M. Principe, *The Secrets of Alchemy*

Encyclopedia Britannica, “Robert Boyle”

Encyclopedia Britannica, “Antoine-Laurent Lavoisier”

Encyclopedia Britannica, “Alchemy”

Encyclopedia Britannica, “Phlogiston”

John Hedley Brooke, *Science and Religion: Some Historical Perspectives*

Peter Harrison, *The Territories of Science and Religion*

Carl Sagan, *The Demon-Haunted World*

Chapter 10: Deep Time and the Death of Sacred Chronology

The Shell in the Mountain

Imagine climbing a mountain and finding a seashell in the rock.

Not lying loosely on the ground. Not carried there by a traveler. Not placed there as decoration. Embedded inside stone, high above the sea, surrounded by layers of rock that seem to have been pressed, hardened, lifted, and worn by ages of time.

To an untrained eye, it is a curiosity.

To a religious imagination, it may be explained by a great flood.

To a geologist, it is evidence.

That shell tells a story. It speaks of an ancient sea. It suggests that the rock around it was once sediment under water. It points to pressure, burial, uplift, erosion, and time. Not a little time. Not the time of one king, one dynasty, one civilization, or one scripture. Time far deeper than ordinary human memory can hold.

For much of history, people did not know how to read that record. They knew family stories, tribal memories, royal histories, religious traditions, genealogies, creation stories, and flood stories. They knew time as human beings experience it: birth, childhood, adulthood, old age, death, ancestors, civilization, and sacred beginnings.

But the Earth was keeping a deeper record.

It was written in stone.

When science learned to read that record, one of religion's most intimate assumptions was shaken: the belief that Earth's history could fit comfortably inside sacred chronology.

The rocks told a different story.

The Earth was not young.

The Earth was ancient beyond imagination.

The Earth Beneath Human Memory

Human beings live short lives.

A child is born, grows, becomes old, and dies. A family may remember grandparents and great-grandparents. A village may remember a flood, a famine, a war, or an earthquake. A nation may preserve stories for centuries. A religion may carry sacred memories for thousands of years.

But the Earth is older than human memory.

Much older.

This was not easy for human beings to understand. For most of history, people knew the Earth through stories, not through geology. Mountains looked permanent. Rivers seemed ancient. The sea appeared eternal. Rocks looked silent. Fossils, when noticed, were often treated as curiosities, signs, tricks of nature, remains of a flood, or mysterious objects placed in stone.

The Earth did not yet have a scientific history.

It had sacred histories, creation stories, flood stories, genealogies, myths, and traditions. Many people believed the world was young because their sacred texts and inherited chronologies seemed to say so. Human history and Earth history were compressed into one narrow timeline.

Then geology began to read the rocks.

And the rocks spoke of vanished seas, ancient forests, buried creatures, hardened sediments, lifted mountains, eroded cliffs, cooled lava, deep layers, and immense time. They spoke of a world that had changed slowly, repeatedly, and naturally, long before human beings appeared.

This was one of the greatest shocks in the history of thought.

The Earth was not a recent stage built for human drama.

The Earth was ancient beyond imagination.

Sacred Time and Human Time

Before geology, many cultures placed Earth's origin inside sacred time.

The world began through divine creation. Land, sea, animals, plants, human beings, and moral order were explained through sacred stories. These stories gave people identity. They told them where they came from, who created them, why life mattered, what duties they had, and how they were connected to the divine.

Sacred time is not ordinary time.

It is meaningful time. It does not merely count years. It gives direction, purpose, and identity. It tells a community that the world is not random. It says the universe began with intention, that human life has a place, and that history is guided by divine will.

This made sacred chronology emotionally powerful.

A young Earth felt intimate. It placed creation close to human history. It allowed the world to feel like a family story. It made the universe seem morally arranged around humanity. The first humans, the first sin, the first ancestors, the first covenant, the first sacred command, and the present community all felt connected.

Geology disrupted this.

It did not merely say that a date was wrong. It stretched time so far that human history became tiny. It placed human beings near the very recent edge of a vast planetary story. It showed that mountains, oceans, continents, fossils, and rock layers had histories older than any scripture, older than any temple, older than any language, older than humanity itself.

This was not only a scientific correction.

It was a psychological displacement.

Copernicus moved humanity away from the center of space.

Geology moved humanity away from the center of time.

Rocks as Records

A rock may look silent, but it can be a record.

Sedimentary layers can reveal ancient environments. Fossils can reveal organisms that once lived and died. Volcanic rocks can show past eruptions. Erosion can show the slow work of water, wind, ice, and gravity. Folded rocks can show pressure and movement. Marine fossils high in mountains can show that land now lifted above sea level was once under water.

The Earth carries its history in its body.

But that history must be read carefully.

Before geology matured, many people tried to fit rocks into familiar sacred narratives. Fossils might be explained as remains of Noah's flood. Rock layers might be forced into a short timeline. Mountains and valleys might be treated as results of dramatic divine events rather than long natural processes.

This was understandable because people interpreted evidence through the worldview they already had.

But the evidence slowly resisted.

The layers were too many.

The processes were too slow.

The fossils were too strange.

The extinct creatures were too numerous.

The order of the rocks was too structured.

The Earth looked too old.

Geology became powerful because it let the rocks correct the story.

This is the same pattern we have seen before.

The telescope corrected the sacred sky.

Anatomy corrected ancient medical authority.

Chemistry corrected magical ideas of matter.

Geology corrected sacred chronology.

Nature was speaking again, not through omens or divine messages, but through evidence.

James Hutton and the Long Work of Nature

James Hutton, the Scottish geologist of the eighteenth century, helped transform the way people understood Earth's history.

He studied rocks, landscapes, erosion, sedimentation, uplift, and geological formations. He came to see Earth not as a recent object shaped mainly by sudden supernatural events, but as a planet shaped by natural processes working over immense periods of time.

Hutton's idea was powerful because it used present processes to understand the past.

Rivers carry sediment today.

Waves erode coasts today.

Volcanoes produce new rock today.

Heat and pressure affect Earth materials today.

Land can be worn down, buried, hardened, lifted, and worn down again.

If these processes operate over enough time, they can produce enormous change.

This required a new imagination.

Human beings are naturally impressed by sudden events. A flood, earthquake, eruption, or landslide captures attention. Slow change is harder to see. A cliff erodes grain by grain. Sediment accumulates layer by layer. A mountain range rises over immense time. A canyon deepens through patient water and gravity.

The old mind often looked for catastrophe and divine drama.

Hutton saw the power of slow process.

This was revolutionary.

It suggested that Earth's great features did not need to be explained by a few miraculous events. They could be explained by natural causes acting over vast time.

No Vestige of a Beginning

Hutton is often remembered for the idea that he saw "no vestige of a beginning, no prospect of an end."

The phrase captures the emotional force of deep time.

He was not saying the Earth had no beginning in the absolute sense. He was saying that when he looked at geological evidence, he saw cycles of formation, erosion, deposition, uplift, and renewal extending far beyond the visible reach of human history.

The Earth's history was not shallow.

It was deep.

This changed the scale of human thought. A few thousand years could not contain the story told by rocks. Sacred chronologies built from genealogies and ancient texts became inadequate. The Earth demanded a much larger timescale.

Deep time is difficult to feel.

We can say millions or billions of years, but the mind does not easily hold such numbers. Human life is measured in decades. Families are measured in generations. Civilizations are measured in centuries or millennia. Geology asks us to think in spans so vast that human history becomes a thin film on the surface of time.

This can be unsettling.

A young Earth makes human beings feel close to the beginning. A deep-time Earth makes us late arrivals.

But truth is not obligated to preserve human comfort.

The rocks did not care whether their age fit sacred expectation.

They had their own record.

Lyell and the Power of Gradual Change

Charles Lyell carried geological thinking further in the nineteenth century.

His great work, *Principles of Geology*, argued that the present is the key to the past. Natural processes now observed on Earth could explain ancient geological formations if given enough time.

Lyell emphasized gradual change over immense periods. Rivers, erosion, sediment, volcanoes, earthquakes, and other natural processes could shape the planet without requiring constant supernatural intervention.

This idea is often associated with uniformitarianism, though the history and meaning of that term are complex. For this book, the basic point is simple:

The Earth can be explained by natural processes operating over deep time.

That idea weakened sacred history.

If valleys, layers, fossils, and mountains could be explained by natural causes, then divine catastrophe became less necessary. If Earth had a long natural history before human beings, then humanity was not the obvious center of the story. If entire worlds of life had appeared and disappeared before us, then creation was not a simple human-centered event.

Lyell's influence extended beyond geology.

Charles Darwin read Lyell. The idea of slow change over vast time helped prepare Darwin's mind for evolution. Natural selection requires time. A young Earth would not provide enough room for the gradual transformation of life.

Geology gave evolution its stage.

Deep time made Darwin possible.

Fossils and Vanished Worlds

Fossils were among the most disturbing pieces of evidence.

They showed that strange creatures had once lived and died. Some resembled modern animals. Others did not. Some were found in layers far below the surface. Some were marine creatures embedded in rocks far from the sea. Some belonged to species that no longer existed.

Imagine again that shell in the mountain.

What does it mean?

A sacred story may say the world was once flooded. A scientific mind asks a different set of questions. Was this rock once under the sea? How did it rise? What layer contains the fossil? What other fossils appear above and below it? What process could explain its position?

The fossil is no longer treated as a symbol.

It becomes evidence.

Extinction was especially troubling.

If God had created each species perfectly, why had so many disappeared? If creation was centered around human beings, why did so many creatures live and die before humanity appeared? If nature was carefully arranged for human purposes, why were there entire worlds of life that humans never saw?

Fossils did not immediately produce one clear interpretation. Different thinkers explained them in different ways. Some tied them to floods. Some saw them as remains of extinct organisms. Some struggled to fit them into theology.

But the evidence accumulated.

The Earth had a long biological past. Life had changed. Species had appeared and disappeared. Entire forms of life had vanished before human beings existed.

This was devastating to a simple sacred history.

It suggested that the world was not created in one recent event with all creatures arranged for human use. It suggested a deeper, older, stranger story.

Geology did not yet provide Darwin's mechanism. But it opened the door.

It showed that life had a history.

And that history was written in stone.

The Flood Explanation Weakens

For a long time, many people tried to explain geological evidence through a great flood.

This made sense within a biblical worldview. If there were marine fossils on land, perhaps floodwaters had placed them there. If rock layers showed destruction, perhaps they were the remains of a global catastrophe. If fossils were buried, perhaps they were victims of divine judgment.

The flood explanation was emotionally and religiously convenient.

It preserved sacred narrative. It kept Earth history inside scripture. It maintained divine action as the central explanation.

But the more geologists studied the evidence, the less adequate the flood explanation became.

The rock record was too complex. Layers showed different environments, different ages, different organisms, repeated processes, and long sequences. Fossils appeared in ordered patterns, not as a simple chaotic mixture from one flood. Mountains, valleys, sedimentary layers, volcanic deposits, and erosional surfaces pointed to many processes over immense time.

A single recent flood could not explain the full evidence.

This is a key moment in the retreat of God as explanation.

A sacred story that once explained the Earth was replaced by geology.

The Earth did not need a global divine catastrophe to explain its structure. It needed time, water, heat, pressure, erosion, sedimentation, uplift, volcanism, and biological history.

Natural causes did the work.

The Earth Without Human Centrality

Deep time changed humanity's self-image.

In a young-Earth view, human beings appear near the beginning. The world is created, and human beings quickly enter the story. The universe feels focused on us. Earth history and human history are closely linked.

In deep time, the picture is very different.

The Earth existed for billions of years before human civilization. Life existed for vast periods before human beings. Dinosaurs lived and disappeared long before us. Continents shifted. Seas advanced and retreated. Mountains rose and wore down. Ice ages came and went. Species appeared and vanished. The planet changed repeatedly without human eyes to witness it.

This is humbling.

It means Earth is not simply the stage for human religion. It is a planet with its own natural history.

Human beings are recent.

Very recent.

This does not make human life meaningless. It makes human life fragile and precious. We are not the center of geological time, but we are the part of Earth that has become aware of time. We can read the rocks. We can reconstruct ancient worlds. We can understand our late arrival.

But the religious claim of cosmic and temporal centrality becomes harder to maintain.

If Earth's story is billions of years long and human beings occupy only the last tiny part of it, then the idea of a world created mainly for us becomes less convincing.

Deep time is not only geology.

It is humility.

Deep Time and the Retreat of Sacred Chronology

Sacred chronology tried to place Earth's history within humanly meaningful time.

Geology shattered that narrow frame.

The age of the Earth could not be derived reliably from genealogies, ancient stories, or theological calculation. It had to be investigated through rocks, layers, fossils, erosion, and later, radiometric dating.

Modern science has shown that Earth is about 4.54 billion years old. That number is difficult to absorb. It is not just older than any scripture. It is older than human language, older than our species, older than mammals as a dominant group, older than flowers, older than forests, older than animals on land, older than complex life.

Such vast time changes everything.

It gives natural processes room to work. It allows mountains to rise and erode. It allows continents to move. It allows species to evolve and vanish. It allows life to become a long, branching history rather than a recent arrangement of fixed forms.

This is why geology was so dangerous to religious literalism.

It did not attack God directly.

It attacked the timeline.

And once the timeline changed, many religious explanations had to change with it.

A young Earth could support a simple creation story.

An ancient Earth demanded a different understanding.

A recent creation could center humanity.

Deep time made humanity late.

A sacred flood could explain some imagination.

Geology required natural processes across ages.

The rocks quietly contradicted the old story.

Natural Explanation Proves Itself

Geology did not win merely because it sounded impressive.

It worked.

It explained why rocks occur in layers. It explained why marine fossils appear far from the sea. It explained how sediment becomes stone. It explained erosion, uplift, volcanoes, earthquakes, mountains, valleys, and ancient environments. Later, with radiometric dating, it gave increasingly reliable ways to estimate the ages of rocks and the Earth itself.

This practical success matters.

A sacred flood story may give moral meaning, but it does not explain the full order of fossils in layered rocks.

A young-Earth chronology may feel comforting to a believer, but it does not fit the evidence of geology.

A divine catastrophe may sound dramatic, but it does not explain the slow accumulation of sediments, the rise and erosion of mountains, the succession of fossils, and the repeated changes in Earth's surface.

Geology explains these things better.

It does not explain everything. No science does. But it gives a natural framework that can be tested, corrected, expanded, and used.

It helps locate minerals, understand earthquakes, study climate history, find water resources, reconstruct ancient environments, and prepare for hazards.

The rocks became useful because they became understandable.

That is the mark of a powerful explanation.

What Deep Time Did to Design

Deep time also changed the argument from design.

Before geology and evolution, living things seemed to appear within a relatively short history. Their complexity seemed immediate and miraculous. Eyes, wings, hands, flowers, instincts, and organs looked like objects directly fashioned for purpose.

But deep time changed the possibilities.

If Earth is very old, then life has had immense time to change. If environments shift over long periods, then organisms face changing pressures. If species appear and disappear across geological layers, then life has a history. If fossils show succession, then living forms are not fixed in one recent creation.

Deep time does not itself prove evolution by natural selection. Darwin would do that more fully.

But deep time made evolution plausible.

It gave nature the one thing gradual change requires: time.

Without deep time, Darwin's theory would have struggled. With deep time, the slow accumulation of small changes became imaginable.

This is another example of how scientific discoveries support one another.

Astronomy displaced Earth from the center of space.

Geology displaced humanity from the center of time.

Evolution would later displace human beings from a separate biological creation.

Each discovery weakened a different form of religious centrality.

God became less necessary not because one argument destroyed belief, but because many lines of evidence slowly changed the whole picture of reality.

Disaster Without Divine Judgment

Geology also changed how people understood disasters.

Earthquakes, volcanoes, floods, landslides, and tsunamis had often been interpreted as signs of divine anger or judgment. If the earth shook, God was displeased. If a city was destroyed, perhaps it was sinful. If a volcano erupted, perhaps a god demanded attention.

Geology moved these events into natural explanation.

Earthquakes come from movements within the Earth. Volcanoes come from molten rock, pressure, tectonic settings, and geological processes. Floods come from rainfall, rivers, topography, storms, and climate. Landslides come from gravity, water, rock structure, and slope failure.

These explanations do not remove tragedy. They do not dry the tears of parents. They do not rebuild homes by themselves. They do not answer every emotional question after disaster.

But they do something morally important.

They remove false blame.

A child crushed in an earthquake is not being punished by God. A village destroyed by a tsunami is not proof of divine anger. A family killed by a volcanic eruption is not evidence of moral failure.

Natural disasters are natural.

That understanding helps us respond with compassion and preparation rather than judgment.

We can build safer structures.

We can map hazards.

We can monitor volcanoes.

We can issue warnings.

We can plan evacuation routes.

We can improve disaster response.

Geology does not make us all-powerful.

But it makes us less superstitious and more responsible.

The Religious Reply

A thoughtful believer may say:

"Deep time does not disprove God. God could have created through long natural processes. The days of creation may be symbolic. Scripture may teach spiritual truth rather than geological science. A billions-year-old Earth can still be God's creation."

This reply is reasonable.

Many religious believers accept modern geology. Many do not read sacred texts as literal science. Many see deep time as part of divine creation rather than a threat to it.

But this reply also shows the pattern again.

Religion adapts after science forces revision.

The older literal timeline becomes harder to defend. The sacred chronology must be reinterpreted. The flood story must be softened, symbolized, localized, or removed from geological explanation. Earth's age is no longer learned from scripture. It is learned from science.

That is the retreat.

The believer may still say that God created the Earth over deep time. But geology explains the Earth's structure and history through natural processes. God is not needed to explain rock layers, fossils, erosion, volcanoes, sedimentation, uplift, or the age of the planet.

God moves behind the process.

Science explains the process.

Evidence-Based Response

The evidence-based response is clear.

Geology revealed an ancient Earth shaped by natural processes over immense time. Hutton and Lyell helped show that the processes we observe today, such as erosion, sedimentation, volcanism, and uplift, can explain much of Earth's past when given enough time. Fossils revealed vanished worlds and the succession of life. Rock layers showed complexity far beyond a single recent catastrophe.

Sacred chronology became inadequate.

This does not logically disprove every idea of God. But it does make God unnecessary as a geological explanation.

Mountains do not need divine lifting.

Fossils do not need to be explained by one sacred flood.

Rock layers do not need supernatural chronology.

Earthquakes do not need divine anger.

Volcanoes do not need gods beneath the mountain.

Earth's age does not need to be calculated from ancient genealogies.

Geology gives better explanations.

It gives evidence, process, timescale, and testable understanding.

That is why deep time was such a major step in the retreat of God.

Chapter Summary

For much of history, Earth's past was understood through sacred stories, creation accounts, flood narratives, and short chronologies. Human history and Earth history were often placed close together.

Geology changed this deeply.

James Hutton and Charles Lyell helped show that Earth was shaped by natural processes working over immense time. Erosion, sedimentation, uplift, volcanism, and other processes could explain much of the planet's structure. Fossils revealed vanished worlds and forms of life that existed long before human beings.

Deep time displaced humanity from the center of Earth history. It weakened sacred chronology and made a young Earth increasingly impossible to defend scientifically.

Geology did not disprove God in every philosophical sense. But it made God unnecessary as an explanation for Earth's structure, age, and natural history.

Key Points

Before geology, many people understood Earth's history through sacred chronology and flood stories.

Hutton helped reveal the power of slow natural processes over immense time.

Lyell emphasized that present processes can help explain the geological past.

Fossils showed that life had a long history and that many species had vanished.

Deep time made human beings recent arrivals in Earth's story.

Geology weakened young-Earth thinking and sacred timelines.

Natural disasters are better explained by geology than by divine anger or punishment.

Deep time prepared the way for Darwin's theory of evolution.

Transition to the Next Chapter

Geology gave science the vast time it needed.

But time alone was not enough. Another question remained.

If the Earth is ancient and fossils show a long history of changing life, how did living things become so complex? Why do organisms appear so well adapted to their environments? Why do eyes, wings, flowers, hands, instincts, and organs look as if they were designed?

For centuries, this apparent design seemed like one of the strongest arguments for God.

Then Darwin gave the world a different answer.

In the next chapter, we will see how life could look designed without being designed.

Sources

James Hutton, *Theory of the Earth*

Charles Lyell, *Principles of Geology*

Martin J. S. Rudwick, *Earth's Deep History*

Martin J. S. Rudwick, *Bursting the Limits of Time*

Stephen Jay Gould, *Time's Arrow, Time's Cycle*

Claude C. Albritton, *The Abyss of Time*

David R. Oldroyd, *Thinking About the Earth*

G. Brent Dalrymple, *The Age of the Earth*

Encyclopedia Britannica, "James Hutton"

Encyclopedia Britannica, "Charles Lyell"

Encyclopedia Britannica, "Geologic Time"

U.S. Geological Survey, "The Geologic Time Scale"

National Academy of Sciences, *Science, Evolution, and Creationism*

John Hedley Brooke, *Science and Religion: Some Historical Perspectives*

Chapter 11: Darwin and Design Without a Designer

The Eye That Seemed to Prove God

Few things in nature look more designed than the eye.

It gathers light. It focuses images. It adjusts to brightness. It detects motion. It helps an animal find food, avoid danger, recognize others, and survive in a world full of threats.

To look at an eye before Darwin was to feel the force of the design argument.

How could such a thing exist without a designer?

The same question could be asked about wings, hands, flowers, lungs, kidneys, instincts, camouflage, nests, spider webs, and the human brain. Everywhere in life, there appeared to be structure serving function. Parts seemed fitted to purpose. Organs seemed matched to need. Animals seemed adapted to their surroundings with astonishing precision.

A bird's wing looked made for flight.

A fish's fin looked made for water.

A flower's color and scent looked made for pollination.

A hand looked made for grasping.

The living world seemed full of intention.

This is why biological design became one of the strongest arguments for God. If a watch suggests a watchmaker, then surely the eye suggests an eye-maker. If a house needs a builder, surely the body needs a divine engineer. If a machine requires design, surely life requires a designer.

This argument felt powerful because it fit ordinary experience.

A chair does not appear without a carpenter. A book does not write itself. A temple does not rise without builders. A clock does not form from scattered metal. Complex things in human life usually come from minds.

So life seemed to point to the greatest mind of all.

Then Darwin changed the question.

What if life could look designed without being designed?

That question shook religion more deeply than almost any scientific question before it.

The World Before Darwin

Before Darwin, many people believed that species were fixed.

A horse was a horse. A bird was a bird. A fish was a fish. A human being was a human being. Each kind of living thing was thought to have been created separately, according to divine plan.

This view fit religious tradition well.

It made nature look ordered, purposeful, and complete. Every creature had its place. Every organ had its purpose. Every species reflected the mind of the creator. The natural world became a kind of living museum of divine wisdom.

This view also fit the limited evidence available to ordinary people.

Most changes in life are small on the scale of a human lifetime. Dogs give birth to dogs. Birds hatch from eggs and become birds. Oak trees produce acorns that grow into oak trees. Children resemble parents. Farmers and breeders could alter animals and plants through selection, but species still seemed basically stable.

The idea that all living things might be related through common descent was not obvious.

It required a larger view of time, variation, inheritance, geography, fossils, extinction, and struggle.

That larger view became possible only after many earlier changes in thought. Geology had revealed deep time. Fossils had shown vanished worlds. Exploration had brought naturalists into contact with unfamiliar plants and animals. Breeders had shown that traits could change across generations. Scientists had collected, compared, classified, and studied living forms in increasing detail.

Darwin did not work in isolation.

He stood at the meeting point of many lines of evidence.

But his central insight was extraordinary.

Nature itself could select.

The Journey That Changed a Mind

Charles Darwin began his voyage on HMS *Beagle* in 1831 as a young naturalist.

He was not yet the Darwin of history. He was curious, observant, patient, and deeply interested in nature. The journey took him to South America, the Galápagos Islands, Australia, and other parts of the world. He saw landscapes, fossils, animals, plants, peoples, and geological formations that expanded his mind.

Travel matters in science.

It removes the illusion that one's local world is the whole world. A person who sees only one landscape may think it is typical. A person who sees many landscapes begins to understand variety. A person who sees only local animals may think life is fixed and simple. A person who sees islands, fossils, and geographical patterns begins to ask deeper questions.

Darwin noticed that living things were not randomly distributed.

Islands had species similar to nearby mainland forms, yet different in important ways. Fossils resembled living animals in the same regions. Closely related species appeared adapted to different environments. Similar organisms showed both continuity and change.

These patterns were difficult to explain if every species had been separately created in its present form.

Why would island species resemble nearby mainland species if each had been independently designed?

Why would extinct animals resemble living animals from the same region?

Why would related species show both similarity and difference, as if they shared a history?

Darwin slowly began to see life not as a set of separate creations, but as a branching history.

Life had a past.

Species had relatives.

The living world had changed.

Natural Selection in Plain Language

Darwin's great idea was natural selection.

The idea can be stated simply.

Living things vary. Individuals in a population are not exactly the same. Some are larger, smaller, faster, slower, stronger, weaker, more resistant, more fertile, more visible, or better camouflaged than others.

Some of these variations can be inherited.

More individuals are born than can survive.

Life involves struggle. Organisms compete for food, space, mates, safety, light, water, and survival. Many die before reproducing.

If certain inherited traits help survival or reproduction in a particular environment, individuals with those traits are more likely to leave offspring.

Over many generations, useful traits become more common.

This is natural selection.

There is no foresight in it. Nature does not plan. Nature does not imagine a future eye, wing, lung, or brain. Nature does not say, “Let me design a better animal.”

Small inherited differences affect survival and reproduction. The environment filters them. Generation after generation, the population changes.

That is all.

And yet, over deep time, this simple process can produce astonishing adaptation.

This is why Darwin was so dangerous to the old design argument.

He did not merely say that life was not designed. He offered a mechanism by which life could become adapted without a designer.

That was the revolution.

Design Without a Designer

The phrase "design without a designer" may sound strange at first.

Design usually implies intention. A designer imagines a goal, makes a plan, chooses materials, and builds toward a purpose. Human design is forward-looking. The watchmaker knows what a watch is before making it. The architect knows the house before drawing it. The engineer knows the bridge before building it.

Natural selection works differently.

It has no future goal.

It preserves what works now.

A slightly sharper vision may help an animal avoid predators or find food. A slightly better wing shape may improve flight. A slightly more useful beak may help a bird eat available seeds. A slightly better camouflage pattern may help an insect survive. A slightly stronger immune response may help an organism live long enough to reproduce.

Each step is judged by survival and reproduction in the present environment.

There is no plan for perfection.

There is only differential survival.

Over time, this can create structures that look designed. But they are not designed in the human sense. They are accumulated results of variation, inheritance, selection, and time.

This is difficult because human beings naturally think in terms of intention. When we see order, we imagine an orderer. When we see complexity, we imagine a mind behind it. When we see function, we imagine purpose.

Darwin showed that this habit can mislead us.

Life can have function without foresight.

Adaptation can arise without intention.

Complexity can grow without a divine engineer.

The End of Special Creation

Special creation is the idea that each species was separately created by God in its present or original form.

Darwin weakened that idea deeply.

If species change over time, then they are not fixed. If different species share common ancestry, then they are not separate divine productions. If life branches like a tree, then the natural world is a history, not a static collection.

This changed the meaning of life's diversity.

The similarities among organisms were no longer simply signs of a common designer using similar patterns. They could be signs of common descent. The differences among species were no longer merely separate acts of creation. They could be the result of adaptation, isolation, inheritance, and time.

The tree of life replaced the ladder of creation.

This was especially disturbing when applied to human beings.

If humans share ancestry with other animals, then humans are not a separate creation outside nature. We are part of life's history. We belong to the biological world. Our bodies, senses, emotions, instincts, and even aspects of our behavior have evolutionary roots.

A later book in this series will examine human evolution in greater depth. Here, we should simply mark the historical turning point.

Darwin made it possible to understand human beings as part of nature.

That was a major blow to religious exceptionalism.

Why Darwin Was More Dangerous Than Copernicus

Copernicus moved Earth away from the center of the cosmos.

Darwin moved human beings into the family of life.

Both were displacements, but Darwin's was more intimate.

Most people can live with the idea that Earth moves around the Sun. It may disturb cosmic pride, but it does not touch the body directly. Evolution touches the body, ancestry, identity, and the meaning of being human.

It says that our hands, eyes, bones, muscles, blood, nerves, and brains are not separate miracles. They are products of natural history.

It says we are related to other living beings.

It says the human body carries marks of ancestry.

It says life is not arranged in a simple hierarchy with humans standing outside and above nature.

For many religious believers, this was harder to accept than heliocentrism.

A moving Earth could be reinterpreted. God could still care about human beings on a moving planet. But evolution challenged the idea that humans were specially created in a separate act. It challenged the idea that every species had a fixed divine essence. It challenged the idea that biological complexity required direct design.

Darwin did not merely change science.

He changed the human self-image.

The Religious Shock

The religious shock of Darwin came from several directions.

First, Darwin challenged the design argument. If natural selection can explain adaptation, then the eye, wing, hand, and flower do not require direct divine construction.

Second, Darwin challenged fixed species. The living world was not static. It changed.

Third, Darwin challenged human uniqueness. Humans were part of nature, not a separate biological category outside the animal world.

Fourth, Darwin challenged the idea of a perfectly designed world. Natural selection is wasteful, harsh, and indifferent. It works through struggle, death, extinction, and differential reproduction. That is very different from the image of a loving designer carefully arranging life for comfort and happiness.

This last point matters greatly.

Nature is not gentle.

Many organisms survive by killing others. Parasites live inside bodies. Predators tear prey apart. Diseases spread. Young animals die. Extinction is common. Most organisms that have ever lived are gone.

This does not look like perfect moral design.

It looks like natural history.

A later book will explore this more fully. For now, it is enough to say that Darwin did not merely explain adaptation. He also made the cruelty and waste of nature more understandable without invoking divine purpose.

Darwin's Own Caution

Darwin was careful.

He knew that his ideas would disturb many people. He delayed publishing for many years. He gathered evidence. He studied breeding, fossils, geography, variation, classification, embryology, and many other fields. He did not rush lightly into controversy.

In 1859, he published *On the Origin of Species*. The book did not fully discuss human evolution in its first edition, though the implication was clear. Later, in *The Descent of Man*, Darwin addressed human ancestry more directly.

Darwin's tone was not the voice of a reckless destroyer. It was the voice of a naturalist building a careful argument.

This matters for our book.

Darwin should not be presented as a man who set out simply to attack God. He was doing science. He was trying to explain the patterns of life.

The religious consequences were enormous, but the work itself was grounded in observation and evidence.

This is the pattern again.

Science often weakens God not by direct attack, but by better explanation.

Darwin did not need to say, “There is no designer,” in order to make design less necessary.

Natural selection did that.

The Power of a Natural Mechanism

The most important word in Darwin’s work is mechanism.

Before Darwin, many people saw adaptation and said, “God designed it.”

Darwin said, in effect, “Here is how adaptation can arise naturally.”

That changed everything.

A mechanism is powerful because it does explanatory work. It shows the process by which something happens. It connects causes to outcomes. It allows evidence to accumulate. It allows testing, refinement, and expansion.

Natural selection explained why organisms fit their environments. It explained why related species resemble one another. It explained why island species often resemble nearby mainland species. It explained why artificial selection by breeders could produce change. It explained why variation mattered. It explained why deep time was necessary.

It did not answer every question immediately. Darwin did not know modern genetics. DNA had not yet been discovered. The full mechanisms of inheritance were not yet understood. Later biology would strengthen and refine evolution enormously.

But Darwin had given science the central natural process.

Life no longer needed separate acts of divine design.

That was the crucial change.

Evolution Does Not Mean Random Chaos

One common misunderstanding is that evolution says life is pure accident.

That is not correct.

Variation may arise without regard to what an organism needs. But natural selection is not pure randomness. It is a filtering process. Certain traits help survival or reproduction in particular environments. Those traits become more common over generations.

This combination of variation and selection can produce order.

It is not planned order.

It is not moral order.

It is not perfect order.

It is not designed order.

But it is real order.

This is why evolution can create the appearance of design. The environment repeatedly filters inherited variation. Over time, organisms become adapted to their conditions.

This point must be explained clearly because many religious objections depend on misunderstanding.

They ask, "How could random chance create an eye?"

But evolution is not simply random chance creating an eye. It is cumulative selection acting over deep time on inherited variation, with each small change preserved only if it helps enough in a particular context.

Again, a later book will explain this in more detail.

Here, the main point is historical: Darwin gave science a natural mechanism strong enough to challenge design.

Imperfect Design

Darwin also helped make sense of imperfect design.

If organisms were created by a perfect designer, we might expect elegance, efficiency, kindness, and flawlessness. But living bodies often look improvised.

The human spine is prone to pain. Wisdom teeth often cause trouble. The appendix can become inflamed. The eye has a blind spot. Childbirth can be dangerous. Many animals have structures that make sense only in light of ancestry, modification, and constraint.

Evolution explains this.

Natural selection does not build from scratch. It modifies what already exists. It works with inherited structures. It produces "good enough" survival, not perfection. It is constrained by history. It cannot simply redesign the body like an engineer starting with a clean sheet of paper.

This makes life look less like perfect design and more like accumulated modification.

A perfect designer is not needed to explain imperfect bodies.

Natural history explains them better.

The New Tree of Life

Before Darwin, life was often imagined as a ladder.

At the bottom were simpler creatures. Higher up were animals. Above them were human beings. Above humanity was the divine. This ladder fit a religious and hierarchical imagination. It made life look arranged from low to high, with humans near the top.

Darwin replaced the ladder with a tree.

Life branches. Species share ancestors. Lineages split. Some branches flourish. Others die. There is no single straight line marching toward humanity. There is no biological ladder with humans as the inevitable crown.

This is one of the most humbling aspects of evolution.

Human beings are not the goal of all life.

We are one branch among many.

We are remarkable, yes. We have language, science, art, memory, moral reflection, and the ability to understand evolution itself. But we are not outside the tree.

We belong to it.

This biological belonging can be unsettling for religious belief. But it can also be morally powerful. If we are part of nature, then we are kin to other living beings. If life is connected, then our responsibility toward the living world becomes deeper, not weaker.

Evolution does not degrade life.

It connects life.

Natural Explanation Proves Itself

Darwin's theory did not remain a clever speculation.

It became the foundation of modern biology.

Later discoveries in genetics, DNA, molecular biology, fossils, comparative anatomy, embryology, biogeography, and population biology strengthened and expanded the evolutionary view. Evolution became not a side idea, but the organizing framework of life science.

This practical success matters.

Evolution helps explain antibiotic resistance. It helps explain viral change. It helps explain why pests develop resistance to pesticides. It helps explain similarities among organisms. It helps guide biology, medicine, agriculture, ecology, and genetics.

A design explanation may sound satisfying at first, but it does not do this work.

"God designed it" does not explain why bacteria become resistant after antibiotic pressure.

"God created each species" does not explain the nested patterns of similarity among living things.

"God made animals for their environments" does not explain the fossil record, shared anatomy, genetic relationships, or the imperfect structures inherited from ancestors.

Evolution explains more.

It predicts more.

It connects more.

It keeps opening new paths of investigation.

That is why it became scientifically powerful.

Darwin did not merely challenge religion. He gave biology a working natural explanation.

Darwin and the Retreat of God

Darwin's place in this book is clear.

Copernicus moved Earth from the center.

Galileo made the heavens physical.

Kepler made the sky mathematical.

Newton united Earth and heaven under law.

Laplace made God unnecessary in celestial mechanics.

Vesalius and Harvey made the body natural.

Boyle and Lavoisier made matter chemical.

Hutton and Lyell made Earth ancient.

Darwin made life historical and natural.

Each step removed a major area from divine explanation.

Darwin's step was especially powerful because life had seemed to be the strongest evidence for God. The living world looked designed. Natural selection showed how apparent design could arise without intention.

This is why Darwin remains so controversial.

He touched the central religious claim that life needs a creator.

If evolution by natural selection explains adaptation, then God is not needed as a biological designer.

A believer may still say that God created the process. A believer may still say that evolution is God's method. But this is no longer the old special-creation argument. It is a retreat to a more distant God.

The direct design explanation has been weakened.

The Religious Reply

A thoughtful believer may say:

"Evolution does not disprove God. God could have used evolution as the method of creation. Natural selection may describe how life changed, but God may still be the source of life, the author of natural law, and the purpose behind the process."

This is a serious reply.

Many religious people accept evolution. Many theologians have tried to reconcile Darwin with belief in God. It is logically possible to believe that God created a universe in which life evolves.

But this reply also changes the original claim.

Before Darwin, many believers argued that biological design directly proved a designer. After Darwin, that argument became weaker. If natural selection explains adaptation, then adaptation no longer requires direct design.

Saying "God used evolution" is different from saying "life cannot be explained without God."

The first is a theological interpretation.

The second is an explanatory claim.

Darwin defeated the second.

Evidence-Based Response

The evidence-based response is this:

Darwin did not prove that no God can exist. But he made God unnecessary as the explanation for biological adaptation and the diversity of life.

Natural selection gives a natural mechanism. Variation, inheritance, struggle, reproduction, and deep time can explain how organisms become adapted to their environments. Common descent explains why living things show patterns of similarity and difference. Fossils, geography, anatomy, embryology, and later genetics strengthened the evolutionary view.

The design argument did not disappear emotionally. Many people still feel that life must have a designer. But the scientific need for that designer was deeply weakened.

Life can look designed without being designed.

That is Darwin's great contribution to the retreat of God.

Chapter Summary

Before Darwin, the complexity and adaptation of life seemed to many people like powerful evidence for divine design. Species were often viewed as fixed creations, each made according to God's plan.

Darwin changed this by offering a natural mechanism: natural selection. Living things vary. Some variations are inherited. More organisms are born than can survive. Traits that help survival and reproduction become more common over generations. Over deep time, this process can produce adaptation and the appearance of design.

Darwin did not prove atheism. He did not answer every question about life. But he showed how life could become complex and adapted without direct divine design.

This was one of the greatest turning points in the history of science.

God became unnecessary as the explanation for biological design.

Key Points

Before Darwin, life's complexity seemed to support belief in a divine designer.

Darwin showed that natural selection could produce adaptation without foresight or intention.

Evolution depends on variation, inheritance, struggle, reproduction, and deep time.

Species are not fixed; they change over time and share common ancestry.

Darwin challenged special creation and human biological exceptionalism.

Evolution does not mean pure random chaos; selection is a natural filtering process.

Darwin did not disprove every idea of God, but he weakened God as the explanation for life's design.

Transition to the Next Chapter

Darwin changed how human beings understood life.

But another source of suffering still seemed to carry deep religious meaning: disease.

For centuries, illness had been explained through sin, demons, curses, impurity, divine punishment, bad air, or fate. Families prayed, priests performed rituals, and communities searched for moral causes while invisible organisms spread from person to person.

Then the microscope, experiment, and germ theory changed medicine.

In the next chapter, we will see how Pasteur, Koch, and others helped explain disease without demons.

Sources

Charles Darwin, *On the Origin of Species*

Charles Darwin, *The Descent of Man*

Alfred Russel Wallace, writings on natural selection

Janet Browne, *Charles Darwin: Voyaging*

Janet Browne, *Charles Darwin: The Power of Place*

Ernst Mayr, *The Growth of Biological Thought*

Peter J. Bowler, *Evolution: The History of an Idea*

Jerry A. Coyne, *Why Evolution Is True*

Richard Dawkins, *The Blind Watchmaker*

National Academy of Sciences, *Science, Evolution, and Creationism*

Encyclopedia Britannica, "Charles Darwin"

Encyclopedia Britannica, "Evolution"

Stanford Encyclopedia of Philosophy, "Evolution"

John Hedley Brooke, *Science and Religion: Some Historical Perspectives*

Daniel C. Dennett, *Darwin's Dangerous Idea*

Chapter 12: Pasteur, Koch, and Disease Without Demons

The Fever in the Child

A child develops fever.

At first, the family hopes it is nothing serious. Perhaps it is only a passing illness. But by evening the child is hotter. His breathing is faster. His lips are dry. His eyes look dull. His mother touches his forehead again and again, as if her hand can measure danger. The family gathers around him in fear.

Now imagine this scene before modern medicine.

There is no thermometer in the home. No blood culture. No chest X-ray. No oxygen monitor. No antibiotic. No intravenous fluid. No vaccine. No intensive care unit. No understanding of bacteria, viruses, immune response, sepsis, dehydration, pneumonia, meningitis, or contaminated water.

The family sees suffering, but not the cause.

What can they do?

They pray. They call a priest or healer. They wonder if someone has sinned. They fear a curse. They ask whether a demon has entered the child's body. They promise offerings. They search for moral meaning. They beg God for mercy.

This was not stupidity.

It was terror in the absence of knowledge.

Few things frightened human beings more than disease. A healthy child could be burning with fever by night. A young mother could survive childbirth, then suddenly develop chills, pain, confusion, and death. A cough could spread through a household. A wound could become red, swollen, foul-smelling, and fatal. A village could watch plague move from one home to another, taking the strong and the weak, the devout and the doubtful, the innocent and the guilty.

Before germ theory, disease seemed mysterious, cruel, and morally disturbing.

People saw fever.

They saw rash.

They saw bleeding.

They saw cough.

They saw diarrhea.

They saw pus.

They saw delirium.

They saw death.

But they could not see bacteria. They could not see viruses. They did not know how contaminated water carried cholera. They did not know how mosquitoes transmitted malaria. They did not know why dirty hands could carry death from one patient to another. They did not know why a small wound could become a fatal infection.

So disease was often given religious meaning.

It was punishment.

It was curse.

It was impurity.

It was demonic attack.

It was divine testing.

It was fate.

It was God's will.

This was understandable. When suffering is terrifying and invisible, human beings search for meaning. A random infection feels unbearable. A punishment, however frightening, at least feels like an explanation. If disease is punishment, perhaps repentance can help. If disease is caused by demons, perhaps exorcism can help. If disease is impurity, perhaps ritual cleansing can help. If disease is divine testing, perhaps endurance has spiritual value.

But there was a serious problem.

The explanation was wrong.

And when the explanation is wrong, people suffer more.

The Cruelty of False Causes

Wrong explanations are not harmless.

If disease is believed to be punishment, the sick may be blamed. If illness is seen as impurity, the patient may be isolated with shame instead of care. If seizures are seen as possession, a suffering person may be feared or abused. If mental illness is seen as demonic, the patient may be denied treatment. If epidemics are seen as divine anger, communities may pray while failing to clean water, control infection, or protect the vulnerable.

As a physician, I see this as one of the most important moral lessons in the history of science.

A false cause can produce false treatment.

A false treatment can cost lives.

Compassion alone is not enough if it is not joined with truth. A family may love a child deeply, but if they believe the fever is caused by a curse, they will not search for the real infection. A village may pray sincerely, but prayer will not identify contaminated water. A priest may comfort the dying, but comfort cannot replace sanitation, vaccination, antibiotics, surgery, sterile technique, and public health.

This does not mean comfort is meaningless.

Patients need comfort. Families need hope. Grief needs tenderness.

But comfort is not cure.

Disease forced humanity to face a hard truth: if we want to reduce suffering, we must understand causes as they truly are.

Germ theory was one of the greatest victories of that truth.

Before Germ Theory

Before germ theory, people explained disease in many ways.

Some blamed bad air, often called miasma. Some blamed imbalance of bodily humors. Some blamed curses, demons, sin, impurity, divine punishment, fate, or moral failure. Some treatments were based on herbs, diet, bleeding, purging, prayer, ritual, isolation, or folk tradition.

Some practices helped by accident. Isolation could reduce spread, even if the reason was misunderstood. Clean water could help, even before microbes were known. Certain herbs had real pharmacologic effects. Practical observation sometimes gave useful results.

But without knowing the true cause of infection, medicine remained limited.

A surgeon could operate skillfully, then lose the patient to infection. A physician could visit one patient after another, unknowingly carrying disease. A mother could die after childbirth because invisible organisms entered her body. A city could suffer cholera because sewage contaminated drinking water. A wound could become fatal because nobody understood microbial invasion.

The tragedy was not only that people died.

It was that many deaths were preventable, but the cause remained invisible.

The invisible is dangerous.

When the cause cannot be seen, imagination often replaces evidence.

Germ theory changed this by revealing that tiny living organisms could cause specific diseases. The enemy was not divine anger. It was not a demon. It was not moral stain.

It was biology.

The Invisible World

The microscope opened a new world.

To the naked eye, a glass of water may look clean. A hand may look clean. A surgical instrument may look clean. A wound may appear minor. Air may look empty. Milk may seem pure. Food may seem safe.

But the invisible world may tell another story.

Tiny organisms may be present. Some harmless. Some useful. Some deadly.

This was difficult for people to accept. Human beings trust their senses. If we cannot see something, we may doubt its importance. Before microscopes and experiments, it was hard to believe that unseen organisms could kill a strong adult, a newborn baby, or an entire village.

Yet that is what microbes can do.

The discovery of the microbial world did more than add a fact to medicine. It changed the moral meaning of disease.

A patient with infection was not cursed.

A child with fever was not being punished.

A mother dying after childbirth was not morally impure.

A plague victim was not necessarily sinful.

A wound infection was not fate.

Disease had natural causes.

And once causes were natural, prevention became possible.

This is one of the greatest differences between supernatural explanation and scientific explanation. Supernatural explanation often produces fear, blame, and ritual. Scientific explanation produces investigation, prevention, and treatment.

Pasteur and the War Against Spontaneous Generation

Louis Pasteur was one of the central figures in the development of germ theory.

He studied fermentation, spoilage, microorganisms, and disease. His work helped show that microbes were not mysterious products of decay, but living organisms that could enter, grow, and produce change.

One of the important debates of his time involved spontaneous generation, the idea that living organisms could arise from nonliving matter under ordinary conditions. If rotting broth produced microbes, perhaps the microbes simply appeared from the material itself.

Pasteur's experiments challenged this idea. His famous swan-neck flask experiments showed that broth could remain free of microbial growth when protected from contamination, even while exposed to air. When dust and microbes were allowed to enter, growth occurred.

This was a powerful lesson.

Life did not simply appear by magic in the broth.

Microbes came from microbes.

This mattered for medicine because it suggested that contamination was real. If microbes could enter from outside and grow, then disease and infection might also be prevented by controlling contamination.

The implications were enormous.

Cleanliness was not merely ritual purity.

Sterility was not symbolic.

Contamination was not spiritual.

It was biological.

Pasteur helped move the invisible world into the world of evidence.

Fermentation, Spoilage, and Disease

Pasteur's early work on fermentation may seem far from religion, but it was part of the same great movement.

Fermentation had long seemed mysterious. Grapes became wine. Milk soured. Food spoiled. Beer formed. Bread rose. These changes were visible, but their causes were hidden.

Pasteur showed that microorganisms played a role in these processes. Fermentation was not magic. Spoilage was not vague corruption. These were biological and chemical processes involving living organisms.

This changed the way people understood material transformation.

The same kind of thinking could be applied to disease. If microbes could transform food and drink, perhaps microbes could also invade bodies and produce illness.

This was not only a scientific advance. It was a change in imagination.

People had to learn that small causes could produce large effects.

A tiny organism could sour milk.

A tiny organism could infect a wound.

A tiny organism could spread through water.

A tiny organism could kill a human being.

This was difficult but liberating.

Once microbes were understood, disease was no longer a mysterious blow from heaven. It became something that could be traced, controlled, and sometimes defeated.

Koch and Specific Causes

Robert Koch gave germ theory another powerful foundation.

He helped show that specific microorganisms cause specific diseases. His work on anthrax, tuberculosis, and cholera was especially important. Koch and his colleagues developed methods for growing, isolating, staining, and identifying bacteria. His famous postulates helped connect a particular microbe with a particular disease.

This was a major step.

It was not enough to say vaguely that germs cause illness. Medicine needed specificity.

Which germ?

Which disease?

How does it spread?

Where does it live?

How can it be identified?

How can it be prevented?

How can it be treated?

Specific causes changed medicine.

Tuberculosis was not a curse. It had a causative organism. Cholera was not divine punishment. It could be linked to contaminated water and a specific bacterium. Anthrax was not mysterious fate. It involved a specific microbe.

Specific causes allow specific action.

This is why science is so powerful. It does not merely replace one vague answer with another vague answer. It sharpens the question. It makes the invisible identifiable. It turns fear into investigation.

A demon cannot be cultured in a laboratory.

A bacterium can.

That difference changed history.

Lister and the Clean Wound

The work of Joseph Lister showed how germ theory could transform surgery.

Before antiseptic surgery, operations were dangerous not only because of pain and bleeding, but because of infection. A surgeon could perform a technically successful operation and still lose the patient days later to sepsis. Hospitals could become places where infection spread. Surgical wounds often became contaminated.

Lister applied the idea that microorganisms caused infection. He used antiseptic methods, especially carbolic acid, to reduce contamination in wounds and surgical settings. Over time, antisepsis and later asepsis changed surgery profoundly.

The meaning is simple but powerful.

If wound infection is caused by invisible organisms, then preventing contamination can save lives.

This is science at its best.

An idea becomes a practice.

A practice reduces suffering.

A patient survives who might have died.

A dangerous operation becomes safer.

A hospital becomes less deadly.

This is what supernatural explanation could not do.

Calling infection a curse did not sterilize instruments. Calling pus a sign of spiritual impurity did not prevent sepsis. Praying over a wound did not identify bacteria. Ritual cleansing did not replace antiseptic technique.

The clean wound was not a theological victory.

It was a scientific one.

Semmelweis and the Tragedy of Ignored Evidence

Ignaz Semmelweis deserves mention because his story is one of the most painful in medical history.

In the nineteenth century, he observed that women giving birth in certain hospital wards were dying at high rates from puerperal fever, also called childbed fever. He noticed that doctors and medical students sometimes came from autopsies to examine laboring women. He required handwashing with chlorinated lime, and deaths dropped dramatically.

He did not yet have full germ theory. He could not explain microbes in the modern way. But he had evidence.

His conclusion was deeply uncomfortable: physicians themselves were carrying deadly material to their patients.

Many rejected him.

This is a sobering lesson.

Science is self-correcting, but not always quickly. Evidence can be resisted when it humiliates authority. Doctors, like priests, kings, and philosophers, can be proud. Institutions can protect habit. Human beings do not like evidence that makes them guilty.

Semmelweis was right before the system was ready to admit it.

His story reminds us that the scientific method is not magic. It depends on human honesty. But it also shows why evidence matters so much. When evidence is ignored, patients die.

Handwashing seems simple now.

But it represented a revolutionary truth: invisible contamination can kill.

Again, disease was not punishment.

It was transmission.

Public Health and the Moral Power of Natural Explanation

Germ theory did not only change individual medicine. It changed public health.

Once disease could be understood as natural transmission, society could act collectively.

Water could be cleaned.

Sewage could be managed.

Food safety could improve.

Milk could be pasteurized.

Hospitals could develop infection control.

Vaccines could prevent disease.

Quarantine could be used more rationally.

Outbreaks could be investigated.

Public health departments could track disease patterns.

This is one of the most important moral achievements of science.

Religion often responded to epidemics with prayer, repentance, ritual, or moral accusation. Sometimes religious communities also provided care, charity, and comfort, and that should be acknowledged. But as explanations, prayer and repentance could not identify contaminated water, infected fleas, respiratory droplets, or bacteria in milk.

Public health required natural causes.

If we understand how disease spreads, we can protect strangers. We can protect children. We can protect the poor. We can protect the elderly. We can protect people who will never know our names.

That is morality made practical.

Science does not make people compassionate automatically. But it gives compassion tools.

A society that understands germ theory can save lives on a scale that supernatural explanations never could.

Vaccination and Prevention

Vaccination is another powerful example.

A vaccine does not work because a priest blesses it. It works because the immune system can be trained. The body can learn to recognize a threat. Immune memory can respond more quickly in the future. Disease can be prevented or made less severe.

This is a natural explanation, but it is astonishing.

Vaccination changed human history. Smallpox, once one of humanity's great killers, was eventually eradicated. Other diseases have been reduced dramatically where vaccination is used effectively.

Think about what this means.

For centuries, people prayed during epidemics. They asked God for mercy. They performed rituals. They searched for sin. They feared divine punishment.

But smallpox was not defeated by prayer.

It was defeated by science, public health, vaccination, and global cooperation.

This does not mean every person involved was nonreligious. Many religious people supported vaccination and public health. But the mechanism of success was natural, not supernatural.

The virus did not retreat because humanity repented.

It retreated because science understood enough to prevent transmission and build immunity.

That is a powerful fact.

Antibiotics and the End of Fatalism

Antibiotics later transformed the treatment of bacterial infection.

Before antibiotics, many bacterial infections were often fatal or severely disabling. Pneumonia, sepsis, wound infections, meningitis, tuberculosis, syphilis, and other diseases could kill despite the best care available at the time.

With antibiotics, medicine gained a new weapon.

Not a perfect weapon. Antibiotics do not work against viruses. They can have side effects. Bacteria can develop resistance. Misuse can cause harm. But used wisely, antibiotics have saved countless lives.

Again, the meaning is clear.

If disease is caused by demons, antibiotics make no sense.

If disease is caused by bacteria, antibiotics can be life-saving.

A correct explanation opens the door to correct treatment.

This is why the retreat of supernatural explanation is not merely intellectual. It is humane. It reduces suffering. It gives families more years together. It allows children to grow. It allows surgery to be safer. It allows childbirth to be less deadly. It allows epidemics to be controlled.

Science does not answer every sorrow. But it has answered many causes of suffering better than religion ever did.

Natural Explanation Proves Itself

Germ theory did not win because it was emotionally comforting.

At first, it was often disturbing. It told people that invisible organisms could contaminate hands, water, wounds, food, milk, and instruments. It told doctors that they themselves might carry death from one patient to another. It told hospitals that some of their own practices were dangerous.

But germ theory worked.

Handwashing reduced deaths.

Antisepsis made surgery safer.

Pasteurization made milk safer.

Clean water reduced cholera and other waterborne diseases.

Vaccination prevented disease.

Antibiotics treated bacterial infections.

Infection control protected patients in hospitals.

Public health saved lives before illness even reached the bedside.

That practical success matters.

A demon explanation did not produce sterilization.

A curse explanation did not produce vaccines.

A divine punishment explanation did not produce antibiotics.

A ritual impurity explanation did not produce modern surgery.

Microbiology did.

That is why disease without demons is such an important chapter in the history of God becoming unnecessary.

The supernatural explanation did not simply lose an argument.

It lost because the natural explanation saved lives.

The Physician's View

In medicine, false explanations are not abstract errors.

They have consequences.

If a patient with meningitis is treated only with prayer, the delay may kill. If a child with pneumonia is denied antibiotics when needed, the child may die. If a diabetic emergency is interpreted as spiritual weakness, treatment may be delayed. If a seizure is treated as possession, a neurological condition may be missed. If a woman with postpartum infection is told she is spiritually impure, she may die from sepsis.

A physician must respect the patient, but also respect reality.

Compassion without truth is not enough. A good intention cannot replace a correct diagnosis. A comforting belief cannot kill bacteria. A sincere prayer cannot sterilize a surgical instrument.

This does not mean the patient's emotional and spiritual life should be mocked. A patient may pray, and that may give comfort. Families may hold religious rituals, and that may help them endure fear and grief.

But medical explanation must remain natural.

The body is natural. Disease is natural. Treatment must address natural causes.

This is not cold.

It is compassionate.

The patient deserves the truth.

Disease Without Blame

One of the greatest gifts of germ theory is that it removed moral blame from many diseases.

A child with tuberculosis was not cursed.

A mother with puerperal fever was not sinful.

A cholera victim was not being punished by God.

A person with wound infection was not spiritually impure.

A village struck by epidemic was not necessarily morally worse than another village.

Disease could be tragic without being deserved.

This is a profound moral advance.

Religious explanations often tried to make suffering meaningful by making it moral. But that can be cruel. It can lead people to ask, “What did you do to deserve this?” Science asks a better question: “What caused this, and how can we help?”

That question is more humane.

It replaces accusation with investigation.

It replaces stigma with treatment.

It replaces fatalism with prevention.

It replaces supernatural fear with practical responsibility.

This does not mean every illness is preventable or every patient can be cured. Medicine has limits. People still suffer. People still die. But accurate explanation allows us to respond with honesty instead of superstition.

The Religious Reply

A thoughtful believer may say:

“Germ theory does not disprove God. God may have created microbes, the immune system, and the laws of biology. Prayer can still comfort the sick. Religious people have built hospitals, cared for patients, and served during epidemics. Science explains how disease spreads, but religion may give meaning during suffering.”

This reply contains important truths.

Religious people have often cared for the sick with courage and compassion. Many hospitals have religious roots. Prayer may comfort some patients. Germ theory does not logically disprove every idea of God.

But again, we must separate comfort from explanation.

When we ask what causes infection, the answer is not demons, sin, or divine punishment. The answer involves microorganisms, host defenses, transmission, environment, hygiene, immunity, and treatment. God is not needed to explain the disease process.

A believer may still find meaning in suffering or comfort in prayer. But the medical explanation belongs to science.

That is the important distinction.

Evidence-Based Response

The evidence-based response is overwhelming.

Germ theory explains disease better than supernatural theories ever did. It identifies causes. It guides prevention. It supports sanitation. It explains infection control. It makes vaccination understandable. It makes antibiotics possible. It reduces death.

Supernatural explanations did not do this.

They often produced fear, blame, ritual, and resignation. Germ theory produced handwashing, antisepsis, sterilization, vaccination, pasteurization, clean water, antibiotics, and public health.

This does not mean science has conquered all disease. It has not. New infections emerge. Antibiotic resistance grows. Viruses remain difficult. Pandemics still occur. Medicine still faces uncertainty.

But the direction is unmistakable.

Disease moved from the realm of demons and divine punishment into the realm of biology.

That movement saved lives.

It also made God unnecessary as an explanation for infectious disease.

Chapter Summary

For much of history, disease was explained through sin, demons, curses, impurity, divine punishment, bad air, or fate. These explanations were emotionally understandable but often medically wrong. They could produce blame, fear, and harmful delay.

Germ theory changed medicine by showing that microorganisms can cause specific diseases. Pasteur helped reveal the microbial world and challenged spontaneous generation. Koch helped connect specific microbes to specific diseases. Lister applied germ theory to surgery through antiseptic methods. Semmelweis showed the life-saving importance of handwashing before germ theory was fully accepted.

The result was one of the greatest transformations in human history. Infection became preventable and treatable in ways earlier generations could not imagine.

Disease no longer needed demons.

It needed diagnosis, prevention, sanitation, vaccination, antisepsis, antibiotics, and public health.

Key Points

Before germ theory, disease was often explained through sin, demons, curses, impurity, divine punishment, miasma, or fate.

False explanations can cause real harm by delaying treatment and blaming the sick.

Pasteur helped show the importance of microorganisms in fermentation, spoilage, contamination, and disease.

Koch helped connect specific microbes with specific diseases.

Lister applied germ theory to surgery through antiseptic practice.

Semmelweis showed that handwashing could dramatically reduce deaths from childbirth fever.

Germ theory transformed public health, vaccination, surgery, and infection control.

Disease became natural, not supernatural.

Transition to the Next Chapter

Science had now moved through the heavens, the planets, the body, matter, Earth, life, and disease.

But the universe itself still remained mysterious.

How large is it? Is it fixed or changing? Are space and time absolute? Is our galaxy the whole universe, or only one small island in a much greater cosmic sea?

The next great transformation came through modern physics and astronomy.

In the next chapter, we will meet Einstein, Hubble, and the vast expanding universe.

Sources

Louis Pasteur, scientific writings on fermentation, spontaneous generation, and germ theory

Robert Koch, scientific writings on bacteriology, anthrax, tuberculosis, and cholera

Joseph Lister, writings on antiseptic surgery

Ignaz Semmelweis, *The Etiology, Concept, and Prophylaxis of Childbed Fever*

Roy Porter, *The Greatest Benefit to Mankind*

Sherwin B. Nuland, *Doctors: The Biography of Medicine*

Jacalyn Duffin, *History of Medicine: A Scandalously Short Introduction*

Paul de Kruif, *Microbe Hunters*

Thomas D. Brock, *Robert Koch: A Life in Medicine and Bacteriology*

Gerald L. Geison, *The Private Science of Louis Pasteur*

Encyclopedia Britannica, "Germ Theory"

Encyclopedia Britannica, "Louis Pasteur"

Encyclopedia Britannica, "Robert Koch"

Centers for Disease Control and Prevention, resources on hand hygiene, vaccination, and infectious disease

World Health Organization, resources on vaccination, sanitation, and infectious disease prevention

Chapter 13: The Universe Beyond God's Old House

The Night Sky Was Larger Than We Imagined

Stand outside on a clear dark night.

Look upward.

At first, the sky may seem still and familiar. A few bright stars. A pale Moon. Perhaps a planet shining steadily. If you are far from city lights, the Milky Way may appear as a faint river of light across the darkness.

For most of human history, that sky looked like a ceiling.

Beautiful, mysterious, unreachable, and sacred.

Ancient people saw gods, ancestors, omens, divine order, and heavenly messages. The sky seemed above the human world, not part of the same physical reality. Even after Copernicus, Galileo, Kepler, Newton, and Laplace had changed astronomy, many people still imagined the universe as far smaller and simpler than it really is.

Then modern physics and astronomy enlarged reality again.

The Earth was not the center.

The Sun was not the center.

The Milky Way was not the whole universe.

Space and time were not simple fixed backgrounds.

The universe was not a static stage.

It was vast, dynamic, lawful, ancient, and expanding.

Albert Einstein changed the meaning of space, time, gravity, and motion. Edwin Hubble helped show that our galaxy is only one among many and that the universe itself is expanding.

Together, their work helped create the modern scientific universe.

This chapter is not meant to fully explain cosmology. That is the task of Book 3. Here, our purpose is more focused.

We need to see how modern physics and astronomy continued the same historical pattern.

The universe became more understandable through natural explanation.

And as it did, God became less necessary as an explanation of the cosmos.

The Older Small Universe

The older religious imagination often pictured a universe that was small, ordered, and human-centered.

The Earth was the main stage. The heavens were above. God, angels, or divine beings occupied the higher realm. Human history mattered cosmically. Human sin, obedience, prayer, sacrifice, and salvation seemed connected to the very structure of reality.

This was emotionally powerful.

A small universe is easier to feel at home in. A human-centered universe flatters us. It tells us that our lives are watched by heaven, that our actions matter to the ruler of all things, and that the cosmos itself is arranged around human destiny.

But astronomy made that view harder and harder to maintain.

The Earth became one planet.

The Sun became one star.

The Milky Way became one galaxy.

Then the Milky Way itself became only one galaxy among many.

The universe kept growing.

This was not only a change in size. It was a change in meaning. The larger the universe became, the harder it was to imagine the old religious cosmos as a literal map of reality.

A God may still be believed in. But the universe no longer looked like the small, layered, human-centered world of ancient imagination.

Science had again forced theology to adjust.

Einstein and the New Meaning of Space and Time

Before Einstein, many people imagined space and time in a simple way.

Space was the container in which things happened.

Time was the universal clock ticking everywhere in the same way.

Gravity was understood through Newton's law as attraction between masses. Newton's physics was extraordinarily successful, and for many practical purposes it still works beautifully. But it was not the final word.

Einstein changed the picture.

His theory of relativity showed that space and time are not simple, separate, absolute backgrounds. They are connected. Motion, gravity, time, light, and matter belong to a deeper physical structure.

In general relativity, gravity is not simply a mysterious pull across empty space. It is related to the curvature of space-time caused by mass and energy.

This idea is difficult, but the basic point can be stated simply.

Matter and energy shape space-time.

The shape of space-time guides the motion of matter and light.

This changed the universe again.

Gravity became less like an invisible pull acting on a fixed stage and more like part of the structure of the stage itself. Space and time were no longer passive. They were physical realities.

This did not make the universe less wonderful.

It made it more astonishing.

But it also made it more natural.

The motion of planets, the bending of light, the behavior of massive objects, and the structure of the universe could be studied through physical law rather than divine management.

A Lawful Universe, But Stranger Than Common Sense

Einstein's universe was lawful, but it was not common sense.

This is one of science's most important lessons.

Reality is not required to match human intuition.

The Earth feels still, but it moves.

The Sun appears to circle us, but it does not.

The heavens look perfect, but they are physical.

The body seems animated by mysterious life-force, but it works through organs, cells, chemistry, and physiology.

Time feels absolute, but it is not so simple.

Space seems empty, but it has structure.

Science repeatedly teaches humility.

Human senses evolved for survival on Earth, not for understanding the whole universe. Our intuition is useful for walking, eating, avoiding danger, recognizing faces, and handling objects. It is not naturally built to understand galaxies, black holes, curved space-time, quantum fields, or cosmic expansion.

This matters for religion because many religious ideas arise from common-sense human imagination.

The sky looks above us, so heaven is imagined above.

The Earth feels central, so humanity feels central.

The body dies, but the person feels inwardly more than flesh, so a soul is imagined.

Events feel meaningful, so divine intention is imagined.

Science often shows that common sense is not enough.

Einstein's work continued that tradition. It did not merely add a new theory. It showed that the universe is stranger, deeper, and more lawful than ordinary human imagination had assumed.

Light Bent, and the World Listened

Einstein's ideas were not accepted merely because they were bold.

They had to meet evidence.

One famous test came during the solar eclipse of 1919. Astronomers observed starlight passing near the Sun and found that the light was bent in a way consistent with Einstein's general relativity. The result became famous and helped bring Einstein worldwide attention.

This moment matters for the theme of this book.

A claim about the universe had to answer to observation.

No priestly declaration could settle it. No sacred tradition could prove it. No emotional preference could decide it. Nature had to be asked.

Would light bend as the theory predicted?

The answer from observation strengthened the theory.

This is how science gains authority. Not by being comforting. Not by being ancient. Not by fitting human pride. It gains authority when reality confirms its predictions.

The bending of light was more than a technical result. It showed again that the universe behaves according to laws that human beings can discover, test, and refine.

The sky was not a canvas of divine symbols.

It was a physical reality, subtle and lawful enough to bend light itself.

The Universe Was Not Static

Einstein's equations had implications for the universe as a whole.

At the time, many scientists believed the universe was static: eternal in its overall structure, neither expanding nor contracting. Einstein himself initially favored a static universe and introduced a cosmological constant into his equations to help maintain that idea.

History later moved in another direction.

The universe was not static.

It was dynamic.

This is important because older religious imagination often pictured creation as a finished structure. The world was made, arranged, and placed in order. The heavens were stable. The cosmic order was set.

Modern cosmology increasingly revealed a universe with history.

Stars are born and die. Galaxies form and change. Space itself expands. The cosmos has a past. It has development. It is not simply a fixed divine architecture.

This does not disprove God. But it changes the conversation.

A dynamic universe is less like a completed temple and more like an evolving physical reality.

It has processes.

It has stages.

It has evidence that can be studied.

The universe becomes part of science, not only theology.

Hubble and the Island Universes

For a long time, people debated the nature of spiral nebulae.

Were they clouds of gas within our own Milky Way?

Or were they separate galaxies, vast systems of stars far beyond our galaxy?

Edwin Hubble helped answer this question. By observing Cepheid variable stars in the Andromeda nebula and elsewhere, he showed that these objects were far too distant to be part of the Milky Way. They were separate galaxies.

This was another massive displacement.

The Milky Way was not the whole universe.

It was one galaxy among many.

Think about the emotional meaning of that.

After Copernicus, Earth was no longer the center of the solar system.

After later astronomy, the Sun was no longer unique.

After Hubble, even our galaxy lost its central status.

Humanity became smaller again.

The religious imagination had already adjusted to a moving Earth. Now it had to adjust to a universe of countless galaxies, vast distances, and cosmic scales beyond ordinary thought.

The universe did not look like a small house built around human beings.

It looked like an immense ocean of galaxies.

The Expanding Universe

Hubble's work also helped establish that distant galaxies are generally moving away from us, and that the farther away they are, the faster they appear to recede. This relationship became known as Hubble's law.

The meaning was extraordinary.

The universe is expanding.

This did not mean galaxies were simply flying through preexisting empty space like sparks from an explosion. The deeper idea is that space itself is expanding, carrying galaxies farther apart on large scales.

This changed the human view of cosmic history.

If the universe is expanding now, then in the past it was denser and hotter. This idea would later develop into Big Bang cosmology, which will be discussed more fully in Book 3.

For this chapter, the main point is simple:

The universe has a natural history.

It changes over time. It can be studied through light, distance, motion, mathematics, and physical law. It is not explained by simply saying, "God made it." That statement gives no mechanism, no prediction, no measurement, no model.

Modern cosmology asks different questions.

How fast is the universe expanding?

How old is it?

What is it made of?

How did galaxies form?

What evidence remains from the early universe?

How do gravity, matter, radiation, and space-time interact?

These questions opened a new scientific understanding of the cosmos.

Light as a Messenger

One of the most beautiful ideas in astronomy is that light carries history.

When we look at the Moon, we see it as it was a little more than a second ago. When we look at the Sun, we see it as it was about eight minutes ago. When we look at distant stars, we may be seeing light that traveled for years, centuries, or thousands of years. When we look at distant galaxies, we may be seeing light that began its journey millions or billions of years ago.

The sky is not only space.

It is time.

This is a profound change from the old sacred sky.

Ancient people looked upward and saw signs, gods, omens, and divine territory. Modern astronomy looks upward and sees history written in light.

Light tells us what stars are made of.

Light tells us how fast galaxies move.

Light tells us whether objects are approaching or receding.

Light tells us about temperature, composition, distance, and age.

The universe becomes readable, not as scripture, but as evidence.

This does not remove wonder. It increases it.

A star is more wonderful, not less, when we understand that its light has crossed immense distances. A galaxy is more magnificent, not less, when we know that it contains billions of stars. The night sky is more powerful, not less, when we realize that we are looking into deep time.

Science does not empty the sky.

It fills the sky with reality.

The End of the Human-Sized Cosmos

The ancient religious cosmos was human-sized in meaning, even when it seemed large to ancient people.

It was built around human concerns. It contained heaven above, Earth below, and divine judgment woven into history. The stars could be imagined as lights, signs, or spiritual symbols. The universe had a moral architecture.

Modern cosmology ended that human-sized cosmos.

The universe is not arranged around our emotional needs.

It is vast beyond comfort. It is ancient beyond scripture. It is dynamic beyond old cosmology. It contains galaxies beyond counting, stars beyond imagination, and distances that dwarf every human empire, temple, scripture, and civilization.

This can feel spiritually disturbing.

If the universe is so vast, are human beings important?

Religion often answers that question by making human beings central to divine concern. Science gives a more humbling answer. We are not physically central. We are not cosmically large. We are not the obvious purpose of billions of galaxies.

But this does not make human life worthless.

Meaning does not require central location.

A child is not valuable because she stands at the center of the universe. She is valuable because she can feel, love, suffer, learn, hope, and live. Human dignity does not depend on cosmic size. It depends on consciousness, experience, relationship, compassion, and responsibility.

Science removes false centrality.

It does not remove value.

Natural Explanation Proves Itself

Modern cosmology is not only grand imagination.

It is evidence-based science.

It uses telescopes, spectroscopy, mathematics, satellites, measurements of cosmic radiation, observations of galaxies, gravitational theory, and increasingly precise instruments. It tests models against what the universe actually shows.

This matters.

A sacred cosmology can give meaning to the sky, but it cannot measure the distance to Andromeda.

A creation story can inspire devotion, but it does not calculate redshift.

A myth can place humanity at the center, but it does not explain the cosmic microwave background.

A theological idea can say the universe has purpose, but it does not tell us how galaxies form, how stars live and die, or how space expands.

Science does that work.

Modern cosmology has launched satellites, predicted cosmic observations, measured background radiation, mapped galaxies, studied black holes, and connected the smallest particles with the largest structures.

That practical and predictive success gives science its authority.

The old sacred sky was meaningful.

The modern universe is measurable.

And when something becomes measurable, supernatural explanation loses much of its old power.

The God of the Expanding Universe

Some believers see the expanding universe as evidence for God.

They argue that if the universe has a beginning, it needs a creator. They may connect Big Bang cosmology with creation. They may say that science has rediscovered what religion always knew: the universe began.

This argument is important, but it belongs mainly in Book 3.

Here, we should only prepare the ground.

The expanding universe does not automatically prove God. A scientific model of cosmic expansion is not the same as a theological creation story. The Big Bang, properly understood, is not simply an explosion in space. It is a model of the universe evolving from an earlier hot, dense state. It is supported by evidence and mathematics, not scripture.

Even if the universe has a beginning in some sense, that does not automatically prove a personal God, a loving God, a prayer-answering God, or the God of any specific religion.

A beginning is not the same as a deity.

A mystery is not the same as proof.

A scientific boundary is not an invitation to insert God too quickly.

Book 3 will examine these arguments carefully. For now, the point is that modern cosmology made the universe a subject of science. It did not return it to ancient religion.

God Moves to the Edges

As science explained more of the universe, God moved again.

In the old view, God could be placed directly above the world, in the sacred heavens, governing the sky, stars, weather, fate, and history.

After astronomy, physics, and cosmology, God became harder to place within the universe.

Not in the clouds.

Not above the Moon.

Not among the planets.

Not in the fixed stars.

Not in the Milky Way.

Not in the gaps between galaxies.

So God moved to the edges of explanation.

God became the creator of the laws.

God became the source of the universe.

God became the reason there is something rather than nothing.

God became the explanation behind the Big Bang.

These are more abstract ideas. Some are philosophically serious and deserve discussion. But they are also evidence of retreat.

God no longer explains the daily motion of the heavens.

God no longer explains the structure of the solar system.

God no longer explains the Milky Way.

God no longer explains galaxies moving away from one another.

Physics and astronomy do that.

God is pushed toward the remaining mystery.

That is the pattern of this book.

Where knowledge grows, God retreats.

The Universe Without Divine Management

Modern cosmology reveals a universe governed by natural processes.

Gravity shapes galaxies. Stars form from gas and dust. Nuclear fusion powers stars. Heavy elements are forged in stars and stellar explosions. Galaxies collide and merge. Black holes form. Space expands. Light carries evidence across cosmic time.

This is not a universe that needs visible divine management at each step.

It is a universe of processes.

A believer may say that God created or sustains those processes. But the processes themselves can be studied without invoking God. The scientific explanation works through physics, not theology.

This matters because religion once claimed the sky as its own.

Heaven was the divine realm. The stars were sacred lights. Celestial events carried messages. The cosmos was the architecture of God's rule.

Modern science changed that.

The sky became astrophysics.

This is one of the greatest transformations in human thought.

The heavens did not become less magnificent.

They became more real.

The Emotional Cost of the Vast Universe

The vast universe can feel lonely.

A small religious cosmos gives comfort. It tells us we are watched, judged, protected, loved, and placed near the center of divine attention. It gives the stars a human meaning. It makes the universe feel morally organized.

Modern cosmology gives no such guarantee.

It shows us a universe of immense scale, long ages, violent events, dead stars, radiation, darkness, black holes, empty space, and galaxies far beyond human reach.

This can produce fear.

But it can also produce maturity.

We do not need the universe to be small in order for life to matter. We do not need Earth to be central in order to love one another. We do not need galaxies to exist for our sake in order to find meaning in truth, compassion, beauty, family, service, art, and understanding.

The universe does not need to flatter us.

Reality is not less worthy because it is not centered on us.

In fact, there is a deeper dignity in accepting the universe as it is.

Science asks us to grow up intellectually.

It asks us to give up cosmic flattery and accept cosmic truth.

That is difficult.

But it is also liberating.

The Religious Reply

A thoughtful believer may say:

"Einstein and Hubble did not disprove God. The vastness and order of the universe may point toward God. The laws of physics may be the laws God created. The expanding universe may even support the idea that the universe had a beginning, and that beginning may require a creator."

This is a serious reply.

Modern cosmology does not prove atheism by itself. Einstein did not disprove God. Hubble did not disprove God. The expanding universe does not automatically settle every philosophical or theological question.

But this reply again moves God away from direct explanation.

God is no longer needed to explain the motion of planets.

God is no longer needed to explain galaxies beyond the Milky Way.

God is no longer needed to explain cosmic expansion as a measured phenomenon.

God is no longer needed to explain light, distance, redshift, gravity, or space-time in scientific terms.

God becomes an interpretation placed behind the scientific model.

That may remain meaningful to a believer, but it is not required by the science.

Evidence-Based Response

The evidence-based response is careful.

Modern physics and astronomy do not answer every ultimate question. They do not tell us why there is something rather than nothing. They do not settle every debate about the origin of the universe. They do not remove all mystery.

But they do explain a great deal.

Einstein changed our understanding of space, time, gravity, and the large-scale structure of reality. Hubble helped show that the universe extends far beyond the Milky Way and that it is expanding. Modern cosmology studies the universe through evidence: light, motion, distance, radiation, mathematics, and physical law.

The old sacred sky has become a natural universe.

God is not needed to explain the evidence that galaxies exist beyond our own. God is not needed to explain cosmic expansion as a measured phenomenon. God is not needed to calculate gravitational effects or interpret the light from distant objects.

The universe may still inspire awe, humility, and wonder.

But awe is not evidence for God.

Wonder is not a scientific mechanism.

Mystery is not proof.

Chapter Summary

Einstein and Hubble helped transform the human understanding of the universe.

Einstein changed the meaning of space, time, gravity, and motion. His work showed that the universe is lawful, but stranger than common sense. Hubble helped show that the Milky Way is not the whole universe and that distant galaxies are moving away from us. The universe became vast, dynamic, ancient, and expanding.

These discoveries did not disprove God in every philosophical sense. But they made the old human-centered religious cosmos impossible to maintain literally. The universe was no longer a small sacred structure arranged around Earth and human beings.

Modern cosmology moved the sky from divine architecture to physical reality.

Once again, God became less necessary as an explanation.

Key Points

Einstein transformed the understanding of space, time, gravity, and motion.

General relativity made gravity part of the structure of space-time.

Hubble helped show that the Milky Way is only one galaxy among many.

Hubble's observations helped establish the expanding universe.

Modern cosmology made the universe vast, dynamic, and lawful.

The old human-centered sacred cosmos became scientifically untenable.

The expanding universe does not automatically prove God.

God may remain a theological interpretation, but God is not needed as the scientific explanation for modern cosmology.

Transition to the Next Chapter

Science had expanded the universe beyond ancient imagination.

But one mystery remained deeply personal.

Why does God feel so real to human beings?

Even when science explains the sky, Earth, body, matter, life, disease, and cosmos, belief does not simply disappear. God remains emotionally powerful. Prayer may still comfort. Ritual may still move the heart. The idea of a heavenly father may still feel deeply real.

The next step in this historical journey is not outward into space, but inward into the human mind.

In the next chapter, we will examine Freud, psychology, and the human need for God.

Sources

Albert Einstein, *Relativity: The Special and the General Theory*

Albert Einstein, "Cosmological Considerations in the General Theory of Relativity"

Edwin Hubble, "A Relation Between Distance and Radial Velocity Among Extra-Galactic Nebulae"

Georges Lemaître, writings on the expanding universe and primeval atom

Alexander Friedman, writings on dynamic cosmological models

Arthur Eddington, *Space, Time and Gravitation*

Steven Weinberg, *The First Three Minutes*

Stephen Hawking, *A Brief History of Time*

Sean Carroll, *From Eternity to Here*

Sean Carroll, *The Big Picture*

NASA, resources on Hubble, galaxies, and the expanding universe

European Space Agency, resources on cosmology and cosmic expansion

Stanford Encyclopedia of Philosophy, "Space and Time: Inertial Frames"

Stanford Encyclopedia of Philosophy, "Cosmology and Theology"

John Hedley Brooke, *Science and Religion: Some Historical Perspectives*

Carl Sagan, *Cosmos*

Chapter 14: Freud, Psychology, and the Human Need for God

The Prayer in the Dark

A person sits alone at night.

The house is quiet. The world outside is dark. Someone he loved has died. Or perhaps he is frightened about his own illness. Perhaps he has failed someone. Perhaps he feels guilty, abandoned, small, or helpless.

He closes his eyes and prays.

No one answers in an audible voice. No visible hand enters the room. No law of nature is suspended. Yet something happens inside him. He feels heard. He feels less alone. He feels as if a presence is near. He feels that someone greater than himself knows his pain.

To that person, God may feel very real.

This is important.

Science can explain the sky, Earth, disease, life, and the universe, but it does not automatically remove the emotional power of God. A person may accept evolution, germ theory, astronomy, and modern medicine, yet still feel God in grief, prayer, guilt, fear, or hope.

This is why belief does not disappear simply because science explains nature.

Religion is not only an explanation of the world outside us. It is also an answer to the world inside us.

Fear.

Guilt.

Love.

Death.

Childhood memory.

Need for protection.

Need for forgiveness.

Need for purpose.

Need for a listener in the dark.

Book 1 examined how human beings created gods through fear, death, nature, dreams, ritual, tribe, and culture. This chapter continues that theme from another direction. It asks how psychology helped explain the emotional power of God-belief.

This is where Sigmund Freud enters the story.

Freud did not explain everything. Many of his ideas are debated. Some are no longer accepted in their original form. But he helped move religion from the sky into the human mind.

That was historically important.

God was no longer only a being above the world.

God became something to be studied within human desire, fear, memory, and need.

Why God Feels So Real

God may feel real for reasons that are deeply human.

God may feel present in grief.

God may feel close during prayer.

God may feel like a father in loneliness.

God may feel like a judge in guilt.

God may feel like comfort near death.

God may feel like meaning when life is painful.

These experiences are powerful. They should not be mocked. A person who prays beside a hospital bed, at a graveside, or during a night of fear is not simply making an argument. He is seeking comfort in the face of human vulnerability.

This is why religion survives where argument alone may weaken.

A scientific explanation of thunder may remove a storm god from the sky, but it does not remove fear from the heart. Germ theory may remove demons from disease, but it does not remove grief when a loved one dies. Evolution may explain life, but it does not automatically give every person courage before death.

Religion lives in this emotional space.

It offers a listening presence.

It offers a cosmic parent.

It offers forgiveness.

It offers hope beyond death.

It offers justice beyond human courts.

It offers meaning when suffering feels unbearable.

This does not prove religion is true. But it explains why religion is so attractive.

A belief can feel true because it meets a deep need.

That is one of the central insights of psychology.

Religion as a Human Need

Long before Freud, thoughtful people understood that religion gave emotional comfort.

A frightened child wants a parent. A grieving widow wants hope. A dying man wants reassurance. A guilty person wants forgiveness. A poor person wants justice. A powerless person wants protection. A suffering person wants meaning.

Religion offers all of these.

It says there is a Father above all fathers.

There is justice beyond human courts.

There is life beyond death.

There is meaning behind suffering.

There is forgiveness after guilt.

There is someone who hears the silent cry.

There is a final order behind the apparent disorder of life.

This is emotionally powerful.

A universe without God can feel cold to many people. It may feel empty, indifferent, and morally silent. It does not promise that suffering will be rewarded, that evil will be punished, that love will survive death, or that someone is watching over the frightened.

Religion fills that emotional space.

This does not prove religion is false. But it does show why religion is psychologically attractive.

A belief that answers deep human needs may feel true because it is needed. But need is not evidence. Hunger does not prove that the food we imagine is real. Fear does not prove that a protector exists. Longing for justice does not prove that cosmic justice is guaranteed. Desire for eternal life does not prove that death is not final.

This distinction is crucial.

A belief may be emotionally necessary and factually unsupported.

Freud made this point forcefully.

Freud and the Father in the Sky

Freud saw religion as connected to childhood helplessness.

Human beings begin life utterly dependent. A baby cannot feed himself, protect himself, understand danger, or survive alone. The child needs parents. The parent becomes the first protector, provider, comforter, rule-giver, and judge.

To a child, the parent is powerful.

The parent can feed, hold, punish, protect, approve, and reject. The child's emotional world forms around this authority. Even when we grow older, some of that early need remains within us. We still want protection. We still fear abandonment. We still long for approval. We still want someone stronger to make the world safe.

Freud believed that God was, in part, an extension of this childhood need.

God becomes the ultimate father.

Stronger than every human father.

Wiser than every human father.

More just than every human father.

Always watching.

Always judging.

Always available in fear.

Always capable of protection.

This is why the image of God as Father is so psychologically powerful.

It gives the adult mind a cosmic parent.

Life is frightening. Nature is indifferent. Disease strikes without fairness. Death comes to everyone. Human parents fail, age, and die. But God the Father remains. God sees. God cares. God can protect. God can forgive. God can reunite the dead.

Freud saw this not as evidence for God, but as evidence of human longing.

In his view, religion was a wish projected onto the universe.

The Future of an Illusion

In *The Future of an Illusion*, Freud argued that religious beliefs are illusions.

By illusion, Freud did not simply mean a foolish error. He meant a belief strongly shaped by human wishes. An illusion may happen to be true or false, but its psychological power comes from desire rather than evidence.

This is an important distinction.

A person may believe in heaven because the evidence is strong, or because he cannot bear the thought that his loved one is gone forever. A person may believe in divine justice because the evidence is strong, or because he cannot tolerate the fact that many cruel people escape punishment. A person may believe in a loving God because the evidence is strong, or because he needs to feel protected.

Freud's challenge was this:

Are religious beliefs accepted because they are supported by evidence, or because they satisfy deep human wishes?

This question remains powerful.

When a grieving person says, "I know I will see my mother again," is that knowledge or longing? When a frightened person says, "God has a plan," is that evidence or comfort? When a guilty person says, "God forgives me," is that external reality or inner emotional relief?

These are sensitive questions. They must be asked with humanity, not cruelty.

Grief is real. Fear is real. Guilt is real. The need for comfort is real.

But emotional reality is not the same as external reality.

A dream may reveal desire. It does not prove that the desired object exists outside the mind.

Freud's importance lies in making religion psychologically understandable without making it factually true.

God as Protection Against Nature

Freud also understood that nature frightens human beings.

Nature gives life, but it also destroys. Storms, earthquakes, disease, hunger, pain, aging, and death remind us that we are not in control. A human being can be intelligent, moral, loving, and hardworking, yet still be crushed by forces that do not care.

Religion softens this terror by personalizing nature.

If nature is controlled by God, then nature is not truly indifferent. Behind the storm there is will. Behind death there is plan. Behind suffering there is meaning. Behind disaster there may be judgment, lesson, or mystery.

This can be comforting because a person is easier to plead with than a process.

You cannot beg a virus for mercy.

You cannot ask an earthquake to reconsider.

You cannot pray to a cancer cell and expect compassion.

You cannot negotiate with gravity.

You cannot persuade time not to age the body.

But you can pray to God.

That is psychologically powerful.

Even if God does not answer in any measurable way, the act of prayer can feel like speaking into a listening universe. It can reduce loneliness. It can organize fear. It can give the suffering person a feeling of connection.

Freud saw this as a continuation of the child's need for a protecting parent.

The adult faces the terrifying forces of nature and recreates a father in the sky.

Again, this does not prove God does not exist. But it explains why belief in God is so emotionally attractive.

The need for protection can create the feeling of presence.

Religion, Guilt, and Obedience

Religion does not only comfort. It also controls.

Freud was deeply interested in guilt, repression, desire, and authority. In religious life, guilt often plays a major role. God watches. God judges. God knows secret thoughts. God sees hidden acts. God punishes disobedience. God demands purity. God rewards submission.

This can shape the inner life powerfully.

A person may feel guilty not only for harmful actions, but for thoughts, doubts, desires, anger, sexuality, or disbelief. A child raised under strong religious authority may internalize the idea that an invisible judge is always watching. Even as an adult, he may feel fear, shame, and anxiety long after he intellectually doubts the belief.

Religion can create moral discipline, but it can also create emotional burden.

It can encourage kindness, charity, humility, and self-control. But it can also produce fear, shame, obedience to authority, and suspicion of natural human feelings.

This dual nature must be acknowledged.

A fair book should not pretend religion only harms. It has comforted millions. It has inspired service. It has helped communities endure suffering. It has given language to grief and hope.

But it has also placed heavy guilt on human beings, especially when normal human emotions are treated as sin, doubt is treated as rebellion, and obedience is treated as holiness.

Psychology helps us understand this.

It shows that religion is not merely about ideas. It enters the emotional structure of the self.

Freud's Limitations

Freud was important, but we should not make him into a final authority.

Some of Freud's theories are debated. Some of his claims were speculative. His views were shaped by his time, his clinical practice, his culture, and his own assumptions. Modern psychology, neuroscience, anthropology, cognitive science, and sociology have developed far beyond Freud.

It would be wrong to build the entire case against religion on Freud.

Religion cannot be reduced only to wish fulfillment. It is also culture, identity, ritual, morality, music, community, memory, power, politics, art, fear, hope, and social belonging. It can serve different functions in different lives.

One person may believe because of childhood training.

Another because of grief.

Another because of community.

Another because of mystical experience.

Another because of philosophical argument.

Another because of fear of death.

Human beings are complicated.

Religion is complicated because human beings are complicated.

Freud gave one powerful lens, not the whole picture.

But his lens remains important because it changed the direction of the question.

Instead of asking only, “Is God real?” Freud asked, “What human need does belief in God satisfy?”

That is a major shift.

Once religion becomes psychologically understandable, its authority changes. The feeling that God is real may be explained by the structure of human need, not by the actual presence of God.

The Mind as a Maker of Gods

The human mind does not passively record reality.

It creates meaning.

It sees patterns. It imagines agents. It remembers selectively. It turns fear into story. It turns grief into hope. It turns guilt into judgment. It turns longing into belief. It turns the dead into ancestors, spirits, or souls. It turns the unknown into intention.

This was central to Book 1, and it remains central here.

The mind is not dishonest by intention. It is trying to survive.

A mind that sees agency quickly may be safer in a dangerous environment. If a noise in the dark might be a predator, it is safer to assume agency and be wrong than to ignore danger and be killed. This habit may help survival, but it also makes human beings prone to seeing invisible agents where none exist.

Psychology and cognitive science help explain why gods are so easy for human minds to imagine.

A god is not hard to think about. A god is like a person, but invisible. A god has knowledge, intention, power, memory, emotion, and will. A god can reward, punish, protect, and hear. The idea fits the social brain very well.

Human beings are experts at thinking about other minds.

We wonder what others know, want, intend, hide, remember, and feel. Religion extends this ordinary social ability into the invisible world. God becomes the ultimate mind watching over all minds.

This makes God psychologically natural.

But psychologically natural does not mean factually true.

Prayer as Inner Conversation

Prayer is one of the strongest reasons God feels real.

In prayer, a person speaks inwardly or aloud to an unseen listener. The act can be deeply emotional. It may bring calm, tears, relief, courage, or a feeling of presence. A person may feel heard even when no external answer arrives.

Psychologically, this is understandable.

Human beings often think in conversation. We speak to ourselves. We imagine what others would say. We remember the voices of parents, teachers, spouses, friends, and lost loved ones. We rehearse arguments. We ask ourselves questions. We answer them internally.

Prayer can use this same inner capacity.

The believer speaks to God and may experience comfort from the act of speaking. The mind may generate reassurance, forgiveness, guidance, or emotional release. This can be real as an experience, even if it does not prove that an external divine being is listening.

This distinction matters.

The comfort of prayer is real.

The conclusion that God exists because prayer feels comforting is not logically secure.

A lonely person may feel less lonely when imagining a loving presence. A grieving person may feel comfort when speaking to a dead loved one. A frightened child may feel safer holding a blanket. The feeling is real. The interpretation may be mistaken.

Psychology does not deny the experience.

It questions the explanation.

Religious Experience and Interpretation

People sometimes report powerful religious experiences.

They may feel presence, light, unity, love, forgiveness, awe, peace, or a sense that God has spoken to them. These experiences can transform lives. They can lead people to change behavior, join religious communities, endure suffering, or devote themselves to service.

We should not mock such experiences.

They are meaningful to those who have them.

But we must also ask how experiences are interpreted.

A Christian may experience Jesus. A Hindu may experience Krishna or Shiva. A Muslim may experience the presence of Allah. A Buddhist may experience enlightenment without a creator God. A spiritual person may experience universal consciousness. A secular person may experience awe in nature or music without interpreting it as God.

The experience may be similar in emotional intensity, but the interpretation often follows culture.

This suggests that religious experience is filtered through expectation, upbringing, language, symbol, and belief.

If God were clearly revealing one universal truth, we might expect more consistency. Instead, religious experiences tend to confirm the traditions people already know.

Psychology helps explain this.

The mind does not experience reality in a pure form. It interprets through memory, culture, fear, desire, and expectation.

That does not prove every religious experience is false. But it does weaken the claim that intense inner experience is reliable evidence of external supernatural reality.

Comfort Is Not Evidence

Religion often survives because it comforts.

That must be stated honestly.

When someone loses a child, the thought of heaven may be the only thing keeping that person from despair. When someone is dying, the belief that death is not the end may bring peace. When someone is lonely, prayer may feel like companionship. When someone feels guilty, divine forgiveness may feel cleansing.

These comforts are powerful.

A humane atheist should not sneer at them.

But compassion requires honesty.

Comfort is not evidence.

A belief may be comforting because it answers a wound. That does not make it true. A dying person may feel comfort believing he will see his mother again. That comfort may be emotionally valuable. But it does not prove an afterlife.

Truth and comfort are not the same thing.

Sometimes they overlap. Often they do not.

Science has no duty to preserve comforting illusions. Its duty is to reality. That can feel harsh. But false comfort has dangers too. It can delay grief, encourage denial, create guilt, support harmful authority, or prevent people from facing life honestly.

The mature question is not, “Does this belief comfort me?”

The mature question is, “Is this belief true?”

Freud pushed religion toward that question.

Natural Explanation Proves Itself

Psychological explanation does not remove the pain of grief or the fear of death.

But it helps explain why religion feels so powerful.

It gives us a natural way to understand prayer, guilt, conversion, religious experience, fear of nature, longing for protection, and the image of God as father. It allows us to study belief as part of human development, culture, memory, emotion, and cognition.

This is important because inner certainty can be misleading.

A person may feel that God is present. Another may feel that an ancestor is near. Another may feel that a spirit entered the room. Another may feel that a dream was a message. Another may feel that a ritual cleansed guilt.

The experiences may be real as experiences.

But psychology asks whether the interpretation is justified.

This is where natural explanation proves itself again.

It does not mock the feeling. It explains how such feelings can arise from the human mind.

That changes the discussion.

If the mind can generate a feeling of presence, the feeling alone cannot prove an external presence.

If grief can generate hope of reunion, the hope alone cannot prove heaven.

If guilt can create a need for forgiveness, the relief of prayer alone cannot prove a forgiving God.

Psychology helps us distinguish emotional power from factual evidence.

That is a major step in understanding religion naturally.

Why Belief Survives Scientific Explanation

By this point in the book, we have seen science explain the sky, the planets, the body, matter, deep time, life, disease, and the expanding universe.

So why does belief remain?

Because religion does more than explain nature.

It gives identity.

It gives community.

It gives ritual.

It gives emotional safety.

It gives moral certainty.

It gives hope after death.

It gives a feeling of being watched, known, forgiven, and protected.

A scientific explanation of lightning may remove Zeus from the storm, but it does not remove grief from the heart. Germ theory may remove demons from disease, but it does not remove fear of death. Evolution may remove special creation, but it does not automatically give people meaning. Cosmology may expand the universe, but it does not hold the hand of a dying person.

Religion survives because it lives in human need.

This is why the retreat of God is uneven.

God retreats from explanation before God retreats from emotion.

A person may no longer need God to explain thunder, but may still need God to endure grief. A person may accept evolution, but still pray when frightened. A person may understand neuroscience, but still feel that a dead loved one is near.

Psychology helps us understand this persistence.

Belief is not only an intellectual conclusion. It is often an emotional shelter.

The Religious Reply

A thoughtful believer may say:

"Freud may explain some human needs, but he does not disprove God. The fact that people need comfort does not mean the comfort is false. Hunger does not prove food is imaginary. The desire for God may exist because God is real. Human longing may point toward its true object."

This is one of the strongest religious replies.

It is true that desire alone does not disprove the object of desire. Hunger corresponds to food. Thirst corresponds to water. Loneliness corresponds to human relationship. Perhaps, the believer says, spiritual longing corresponds to God.

This argument deserves respect.

But it has a weakness.

Not every desire corresponds to reality. The desire to live forever does not prove immortality. The desire for perfect justice does not prove cosmic justice. The desire to see a dead loved one again does not prove reunion after death. The desire to be watched over does not prove a heavenly watcher.

Human beings can desire things that do not exist.

Psychology cannot disprove God by itself. But it can explain why belief in God would arise even if God does not exist.

That is its power.

Evidence-Based Response

The evidence-based response is careful.

Freud did not prove that God does not exist. Modern psychology does not prove atheism by itself. But psychology makes God-belief more understandable as a human product.

God-belief fits many human needs: protection, comfort, justice, forgiveness, control, meaning, and hope after death. The image of God as father fits childhood dependency and the adult need for safety. Prayer fits the mind's capacity for inner conversation. Religious experience is often shaped by culture and expectation. Guilt and obedience can be internalized through religious authority.

These facts do not logically eliminate God, but they weaken the argument that God must be real because God feels real.

The feeling of reality can arise from the mind.

That is the central point.

Psychology moves part of religion from heaven into human nature. It shows that belief may be powerful not because God exists, but because human beings need what God represents.

Chapter Summary

Freud and later psychological thinking helped shift the study of religion inward.

Instead of asking only whether God exists, Freud asked what human needs belief in God satisfies. He saw religion as connected to childhood helplessness, father-longing, fear of nature, guilt, wish fulfillment, and the need for protection.

Freud was not right about everything, and religion cannot be reduced entirely to his theories. But his work helped make religion psychologically understandable. Belief in God can be studied as a human response to fear, suffering, guilt, grief, loneliness, and death.

This does not prove God is false. But it shows that the feeling that God is real may come from human need rather than divine presence.

God became less necessary as an explanation of the outer world. Psychology now made God less necessary as an explanation of parts of the inner world.

Key Points

God often feels real because God answers deep human needs.

Freud viewed religion partly as wish fulfillment and a projection of the father figure.

Religion offers protection against fear, nature, guilt, loneliness, suffering, and death.

Freud's ideas are historically important but should not be treated as final or complete.

Psychology helps explain why prayer, ritual, and religious experience feel powerful.

Comfort is not the same as evidence.

The fact that belief is psychologically natural does not prove that the belief is true.

Transition to the Next Chapter

Psychology helped explain why God feels emotionally necessary.

But the next step goes deeper.

If belief, memory, personality, fear, guilt, vision, voice, and religious experience are connected to the mind, then what is the mind connected to?

Modern neuroscience gives a powerful answer: the brain.

In the next chapter, we will examine neuroscience and the natural mind.

Sources

Sigmund Freud, *The Future of an Illusion*

Sigmund Freud, *Totem and Taboo*

Sigmund Freud, *Civilization and Its Discontents*

William James, *The Varieties of Religious Experience*

Pascal Boyer, *Religion Explained*

Scott Atran, *In Gods We Trust*

Daniel Dennett, *Breaking the Spell*

Michael Shermer, *The Believing Brain*

Justin L. Barrett, *Why Would Anyone Believe in God?*

Ara Norenzayan, *Big Gods*

Robert Sapolsky, *Behave*

V. S. Ramachandran, *Phantoms in the Brain*

John Hedley Brooke, *Science and Religion: Some Historical Perspectives*

Stanford Encyclopedia of Philosophy, "Philosophy of Religion"

Internet Encyclopedia of Philosophy, "Freud's Philosophy of Religion"

Chapter 15: Neuroscience and the Natural Mind

The Stroke That Changed a Person

A man wakes up one morning and tries to speak.

The words do not come.

He knows his wife. He knows the room. He understands that something is terribly wrong. But when he tries to form a sentence, only broken sounds emerge. His intention remains. His fear remains. His awareness remains. But language has partly disappeared.

A woman who once managed her home, remembered birthdays, handled money, cooked for her family, and told stories with warmth slowly begins to forget. First, she misplaces objects. Then she repeats questions. Then names fade. Later, even familiar faces become uncertain. Her family says with pain, "She is still here, but she is not the same."

A patient receives anesthesia before surgery. A medicine enters the bloodstream. Brain activity changes. Consciousness fades. Pain disappears. Memory stops forming. Later, when the medicine wears off, awareness returns.

These are ordinary medical facts.

But they carry extraordinary philosophical weight.

If the mind were a separate, independent soul, why should damage to brain tissue remove speech? Why should brain degeneration erase memory? Why should a chemical change consciousness? Why should an injury alter personality, judgment, emotion, or behavior?

The answer from neuroscience is increasingly clear.

The mind is deeply dependent on the brain.

This does not mean neuroscience has solved every mystery of consciousness. It has not. The subjective experience of being aware remains one of the hardest problems in science and philosophy. But one fact has become very difficult to deny:

When the brain changes, the mind changes.

That fact has transformed the human understanding of the self.

The Most Personal Mystery

The sky can become astronomy.

Matter can become chemistry.

Disease can become microbiology.

Earth can become geology.

Life can become evolution.

But the mind feels different.

The mind is not something we only observe from outside. We live inside it. We do not merely have thoughts. In a deep sense, we feel that we are our thoughts, memories, emotions, hopes, fears, choices, and private inner life.

This is why the mind has always seemed close to the soul.

When a person says, “I,” he does not usually mean his liver, kidney, lung, or spleen. He means the inner self. The one who remembers childhood. The one who loves. The one who feels guilt. The one who hopes after death. The one who prays in silence. The one who fears nonexistence.

For thousands of years, this inner self was often understood as something beyond the body. The body could age, weaken, become sick, and die. But perhaps the self was different. Perhaps the real person was a soul using the body for a while. Perhaps death was not the end because the soul could leave the body and continue.

This idea is emotionally powerful.

It comforts grief.

It softens death.

It promises reunion.

It gives dignity to the inner life.

It says the person is more than flesh.

But modern neuroscience has made the old separation between mind and body much harder to defend.

The mind does not float above the body like a ghost. The mind is deeply tied to the brain.

That fact does not make human beings less precious.

It makes the person more natural.

The Brain Was Once Hidden

The brain is not easy to understand.

It is sealed inside the skull. It has no obvious movement like the heart. It does not expand and contract like the lungs. It does not digest like the stomach. To the naked eye, it looks like soft tissue. For a long time, its true importance was not fully understood.

Ancient people knew that blows to the head could cause unconsciousness, confusion, weakness, or death. Some physicians and philosophers suspected that the brain mattered greatly. Others gave more importance to the heart as the center of life, emotion, and self.

It is easy to understand why.

The heart beats. The chest tightens with fear. The pulse races with love, anger, or danger. Grief feels heavy in the chest. Courage is still called "heart." A kind person has a "good heart."

The heart announces itself.

The brain is silent.

Yet the silent organ inside the skull shapes the person more deeply than any other organ.

It receives the world through the senses. It stores memory. It produces language. It coordinates movement. It regulates emotion. It allows planning, imagination, self-control, recognition, fear, pleasure, belief, and doubt.

The more science studied the brain, the more the soul-like self became brain-dependent.

This did not happen through one discovery. It came from many lines of evidence: brain injury, stroke, epilepsy, tumors, infections, dementia, psychiatric illness, drugs, anesthesia, imaging, electrical stimulation, and modern neuroscience.

Together, they point in one direction.

Mind depends on brain.

When Injury Changes the Person

One of the most powerful forms of evidence comes from brain injury.

A person may be kind, patient, gentle, and careful. Then a head injury damages part of the brain, and his personality changes. He becomes impulsive, angry, careless, inappropriate, or unable to control behavior.

Another person may speak normally for years. Then a stroke damages a region of the brain, and suddenly words disappear. She knows what she wants to say, but cannot say it. Or she can speak fluently, but the words no longer make sense.

A person may recognize objects, but not faces.

A person may see but not understand what is seen.

A person may move one side of the body, but not the other.

A person may lose memory while keeping intelligence in other areas.

A person may become unable to plan, judge, inhibit impulses, or recognize danger.

These facts are difficult for the old soul theory.

If the real self is an independent soul, why does damage to brain tissue alter speech, memory, judgment, behavior, and personality?

Why should a soul lose nouns after a stroke?

Why should a soul forget children after dementia?

Why should a soul become impulsive after frontal lobe damage?

Why should a soul fail to recognize a face when a specific brain network is injured?

The simplest explanation is not that the soul is damaged.

The simplest explanation is that the mind depends on the brain.

This does not solve every mystery of consciousness. But it does make a separate supernatural self less necessary as an explanation of ordinary mental life.

Memory and Identity

Memory is central to identity.

We are not only bodies moving through time. We are also stories. We remember names, faces, homes, voices, injuries, joys, humiliations, friendships, parents, children, failures, and hopes. Memory connects the child we were to the adult we are.

But memory is fragile.

It can be damaged by injury, dementia, infection, lack of oxygen, stroke, seizure, drugs, alcohol, trauma, and degenerative disease. It can be distorted by emotion and suggestion. It can fade. It can be partly invented. It can become unreliable while still feeling certain.

This is deeply important.

Many people imagine the soul as the keeper of identity. But if identity depends on memory, and memory depends on brain function, then the self is far more physical than religion often suggests.

A patient with Alzheimer's disease may gradually lose the names of loved ones, the memory of home, the ability to handle daily life, and eventually even the sense of personal continuity. The family may say, with heartbreak, "He is still here, but he is not the same person."

That sentence carries enormous philosophical weight.

The person is not simply trapped inside an unchanged soul. The person's mind is being altered as the brain is altered.

As a physician, one sees this not as an abstract argument, but as a human tragedy. Dementia does not merely damage memory. It can change the person's world, relationships, judgment, language, mood, and identity.

This is not divine mystery in the old sense.

It is brain disease.

And recognizing it as brain disease makes care more honest, more compassionate, and more useful.

Personality Is Not Beyond Biology

Many people think of personality as something spiritual.

A person is kind, cruel, anxious, calm, generous, suspicious, disciplined, impulsive, cheerful, or fearful. These traits feel like the deep self. They feel like character, soul, or moral essence.

But personality is also linked to the brain.

Brain injury can change it. Dementia can change it. Tumors can change it. Drugs can change it. Hormones can influence it. Sleep deprivation can alter it. Depression can darken it. Mania can accelerate it. Addiction can distort it. Trauma can reshape it.

This does not mean people are nothing but chemicals. That phrase is too crude. Human beings are biological, psychological, social, historical, and moral creatures. But biology is part of the person, not separate from the person.

Religious thinking often treats character as if it floats above the body. A person is righteous or sinful, obedient or rebellious, pure or impure, faithful or faithless. But neuroscience and psychology show that behavior often has physical, developmental, emotional, and environmental roots.

This should make us more humble.

A violent person may still be responsible for harm, but we should also ask about brain injury, substance use, trauma, mental illness, and social conditions. A depressed person is not simply weak. An anxious person is not simply faithless. A person with addiction is not simply morally defective. A child with attention problems is not simply disobedient.

A natural understanding of the mind can deepen compassion.

It moves us from condemnation to understanding.

Seizures, Visions, and Voices

For much of history, seizures were frightening and mysterious.

A person might suddenly fall, shake, stare, lose awareness, speak strangely, see visions, hear voices, or behave in ways that looked otherworldly. Without neuroscience, such events were often interpreted as possession, divine contact, prophecy, curse, or madness.

Today, we understand many such experiences in neurological terms.

Epilepsy involves abnormal electrical activity in the brain. Depending on the brain region involved, seizures can affect movement, sensation, awareness, emotion, memory, smell, vision, hearing, and behavior. Some seizures may produce intense fear, déjà vu, strange sensations, or religious-like feelings.

This does not mean every religious experience is epilepsy. That would be a careless overclaim.

But it does show something important.

Powerful inner experiences can arise from the brain.

A voice can feel external and still come from within the mind. A vision can feel sacred and still have a neurological basis. A feeling of presence can be intense and still be generated by brain processes. A sudden certainty can feel divine and still be human.

This matters because many religions were shaped by visions, voices, revelations, trances, dreams, and ecstatic states.

The existence of such experiences does not automatically prove supernatural reality.

Neuroscience asks a different question:

What is happening in the brain when the experience occurs?

That question moves the discussion from heaven to biology.

Drugs, Anesthesia, and the Chemistry of Consciousness

Few things challenge the idea of an independent soul more directly than anesthesia.

A medicine enters the body. It changes brain function. Consciousness disappears. The person does not merely fall asleep in the ordinary sense. Under proper anesthesia, awareness, memory, pain perception, and response can be profoundly altered.

Then the medicine wears off, and consciousness returns.

This is remarkable.

If consciousness were fully independent of the body, why would a chemical agent turn it off and on?

Alcohol can change judgment.

Sedatives can reduce awareness.

Stimulants can increase alertness.

Hallucinogens can produce visions.

Antidepressants can improve mood in some patients.

Antipsychotic medications can reduce hallucinations and delusions.

Pain medicines can change suffering.

Anesthesia can suspend conscious experience.

These facts do not solve the entire mystery of consciousness, but they show that mental life is deeply dependent on brain chemistry and physiology.

The religious idea of the soul often treats consciousness as something separate from matter. But medicine shows that matter can alter consciousness in predictable ways.

A molecule can change mood.

A gas can produce unconsciousness.

A toxin can cause confusion.

A lack of oxygen can damage memory.

A hormone imbalance can affect emotion.

This does not make human life less meaningful.

It makes the mind natural.

Religious Experience and the Brain

Religious experiences can be powerful.

People may report a feeling of divine presence, unity with the universe, overwhelming love, peace, light, forgiveness, or communication from God. Such experiences may change lives. They may reduce fear, inspire service, or strengthen faith.

A serious book should not mock these experiences.

They are real experiences.

But the interpretation of an experience is not the same as the experience itself.

Neuroscience and psychology ask what happens in the brain when people pray, meditate, chant, fast, enter trance states, hear voices, feel presence, or experience awe. These states may involve attention, emotion, memory, expectation, sensory processing, social cognition, and brain networks related to self and meaning.

Again, we must avoid overclaiming. There is no single "God spot" in the brain that explains all religion. Religious experience is complex. It involves many brain systems and many cultural interpretations.

But the broad lesson remains:

The brain can generate experiences that feel deeply spiritual.

This weakens the claim that a powerful religious feeling is proof of God. It may be proof that the human brain is capable of profound states of awe, comfort, unity, and conviction.

The experience may be transformative.

But transformation is not verification.

The Feeling of a Hidden Presence

Many people have felt a presence when alone.

In grief, a widow may feel her dead husband nearby. In fear, a child may feel watched. In prayer, a believer may feel God listening. In danger, a person may feel guided or protected. In certain neurological or psychological states, a person may strongly sense another being nearby even when no one is there.

This feeling can be extremely convincing.

The human brain is built for social awareness. We are constantly tracking other minds. Who is near me? Who is watching? What does he want? Is she angry? Is someone behind me? Am I safe? Am I judged? Am I loved?

This social brain helps us survive in groups. But it can also create false positives.

We may feel agency where there is none. We may sense intention in random events. We may imagine watchers, spirits, ancestors, demons, angels, or gods.

Book 1 discussed how the pattern-seeking and agency-seeking brain helped create belief in invisible beings. Neuroscience adds another layer: these tendencies are grounded in brain function.

God may feel like a person because the brain is designed to think about persons.

An invisible God may feel present because the mind can generate presence.

That does not prove God is false by itself. But it shows that the feeling of presence can have a natural explanation.

The Self Is Not Simple

Many religious traditions imagine the self as a simple soul.

One person. One inner essence. One spiritual identity that remains the same through life and perhaps survives death.

Neuroscience shows that the self is not so simple.

The self is built from many functions: memory, body awareness, emotion, language, social identity, attention, planning, moral judgment, and continuity over time. These functions can be separated, damaged, altered, or distorted.

A person may lose memory but keep personality for a while. Another may keep memory but lose emotional control. Another may be aware of the world but unable to move. Another may deny ownership of a paralyzed limb. Another may have conflicting impulses. Another may speak without understanding. Another may change after brain injury in ways the family barely recognizes.

The self is not a single glowing object inside the body.

It is more like a living process.

This does not make the self unreal. A process can be real. A flame is a process. A song is a process. A storm is a process. Life itself is a process.

The self is real, but it appears to be brain-based, embodied, and dependent on ongoing biological activity.

This is very different from the old idea of a separable soul.

Natural Explanation Proves Itself

Neuroscience does not only make philosophical claims.

It works in medicine.

When a patient has a stroke, brain imaging can locate the injury. When a patient has seizures, EEG and clinical evaluation can help identify abnormal brain activity. When a patient has a brain tumor, imaging can reveal it and surgery or radiation may treat it. When a patient has Parkinson's disease, medications that affect dopamine pathways can improve movement. When anesthesia is given properly, consciousness and pain response can be controlled during surgery.

This practical success matters.

If seizures are possession, neurological treatment makes little sense.

If depression is only moral weakness, psychiatric care is easily dismissed.

If dementia is only spiritual mystery, brain disease is ignored.

If personality is completely separate from the brain, brain injury should not change behavior.

But medicine shows otherwise.

Natural explanation guides diagnosis, treatment, rehabilitation, prevention, and compassion. It does not answer every question about the soul. It does not remove all suffering. It does not cure every brain disease. But it gives us a working path that supernatural explanation never gave.

A correct understanding of the brain helps patients.

That is the practical proof of its power.

Death and the Brain

Death is where the soul idea becomes most emotionally powerful.

When someone we love dies, it is almost unbearable to think that the person is gone forever. The body is still there for a short time, but the person is absent. The face remains, but the expression is gone. The eyes remain, but no one looks out. The voice is silent. The warmth fades. The hand no longer responds.

It is natural to imagine that something has left.

The soul has departed.

The spirit has gone.

The person has moved to another world.

This belief is understandable. It gives comfort. It protects the heart from finality.

But neuroscience gives a different interpretation.

When the brain permanently stops functioning, the processes that produce consciousness, memory, personality, awareness, and experience also stop. The person is gone not because a ghost has flown away, but because the biological system that sustained the mind has ceased to function.

This is painful.

It may be the hardest truth.

But difficulty does not make it false.

A later book in this series will examine the soul and consciousness more deeply. Here, the historical point is enough: the more we learn about the brain, the less we need the soul to explain the mind.

The mind appears to be what the brain does.

When the brain is damaged, the mind is damaged.

When the brain is altered, the mind is altered.

When the brain dies, the mind appears to end.

This is not emotionally easy.

But it is the direction of evidence.

The Religious Reply

A thoughtful believer may say:

"Neuroscience shows that the brain is involved in mental life, but it does not prove there is no soul. Perhaps the brain is an instrument used by the soul, like a radio receiving a signal. If the radio is damaged, the sound changes, but the broadcast still exists. Brain damage may affect expression of the soul without destroying the soul itself."

This is one of the strongest religious replies.

It deserves respect because it acknowledges the evidence of brain dependence while trying to preserve the soul.

But the analogy has a serious weakness.

In the case of a radio, we have independent evidence of broadcast signals. We can detect them, measure them, transmit them, receive them on another radio, and explain how the system works. For the soul, we do not have comparable evidence. We do not have a measurable soul signal. We do not have a known mechanism by which a nonphysical soul interacts with brain tissue. We do not have reliable evidence that memory, personality, or consciousness continues after brain death.

The radio analogy is emotionally appealing, but it does not solve the scientific problem.

The evidence we do have shows that mental life changes when the brain changes.

That is what any natural theory of mind would predict.

Evidence-Based Response

The evidence-based response is careful but strong.

Neuroscience does not yet explain everything about consciousness. The subjective experience of being aware remains one of the deepest problems in science and philosophy. We should not pretend that every mystery has been solved.

But the evidence overwhelmingly links mind to brain.

Brain injury can change personality. Stroke can alter speech. Dementia can erase memory. Epilepsy can produce visions, fear, or altered awareness. Drugs and anesthesia can change consciousness. Brain disease can affect judgment, emotion, movement, and identity.

These facts make the supernatural soul less necessary as an explanation.

The mind may feel separate from the body, but feeling is not proof. The Sun feels as if it moves around Earth. Earth feels still. The heavens once looked perfect. Many appearances mislead us.

The inner feeling of a soul may be another powerful appearance.

Neuroscience does not remove the dignity of the person. It deepens our responsibility to care for the brain, the body, and the suffering mind. It helps us treat mental illness more humanely. It helps us understand patients without blaming them spiritually.

The mind is natural.

That is not a loss of dignity.

It is a gain in truth.

Chapter Summary

For centuries, the mind seemed to point toward the soul. Memory, personality, consciousness, religious experience, visions, voices, and the feeling of self appeared to be more than bodily processes.

Modern neuroscience has changed this view.

Brain injury, stroke, dementia, epilepsy, drugs, anesthesia, and psychiatric illness show that mental life is deeply dependent on brain function. When the brain changes, the person can change. Speech, memory, judgment, emotion, personality, and awareness can all be altered by physical changes in the brain.

This does not answer every mystery of consciousness, and it does not prove atheism by itself. But it makes a supernatural soul less necessary as an explanation of the mind.

God and soul retreat from the inner world just as divine explanation retreated from the sky, Earth, body, matter, life, disease, and cosmos.

Key Points

The mind feels deeply personal, which is why many cultures imagined a soul.

Neuroscience shows that memory, personality, language, emotion, and identity depend strongly on the brain.

Brain injury and disease can change the person in profound ways.

Seizures, visions, voices, and altered states can arise from brain activity.

Drugs and anesthesia show that consciousness can be changed by physical substances.

Religious experience may be real as experience, but that does not prove a supernatural source.

The soul is less necessary when the mind is understood as brain-based.

Transition to the Next Chapter

We have now followed a long historical path.

God once explained the sky, disease, the body, matter, Earth, life, the universe, and the mind. One by one, science gave better explanations. Not complete explanations for everything, but reliable natural explanations wherever evidence was strong.

The next chapter will bring this whole journey together.

God did not disappear in one argument. God did not fall in one battle. God became unnecessary step by step, as reality became more understandable.

In the conclusion, we will examine the shrinking space for God.

Sources

Antonio Damasio, *Descartes' Error*

Antonio Damasio, *Self Comes to Mind*

Oliver Sacks, *The Man Who Mistook His Wife for a Hat*

Oliver Sacks, *An Anthropologist on Mars*

V. S. Ramachandran, *Phantoms in the Brain*

Michael Gazzaniga, *The Ethical Brain*

Michael Gazzaniga, *Who's in Charge?*

Eric Kandel, *In Search of Memory*

Joseph LeDoux, *The Emotional Brain*

Robert Sapolsky, *Behave*

Patricia Churchland, *Neurophilosophy*

Patricia Churchland, *Braintrust*

William James, *The Varieties of Religious Experience*

Andrew Newberg, *Why God Won't Go Away*

Daniel Dennett, *Consciousness Explained*

Pascal Boyer, *Religion Explained*

Stanford Encyclopedia of Philosophy, "Consciousness"

Stanford Encyclopedia of Philosophy, "Neuroscience"

National Institute of Neurological Disorders and Stroke, resources on brain injury, stroke, epilepsy, and dementia

Conclusion: The Shrinking Space for God

God Did Not Disappear in One Blow

God did not disappear from human explanation in one sudden moment.

There was no single day when science rose and religion fell. There was no one equation, one telescope, one microscope, one fossil, one laboratory experiment, or one book that removed God from the human mind.

The change was slower, deeper, and more powerful.

God became unnecessary step by step.

First, God became unnecessary in the sky. Thunder was no longer the voice of divine anger. Lightning was no longer the weapon of a god. Eclipses were no longer cosmic warnings. Comets were no longer messages from heaven. The Sun, Moon, planets, and stars became objects of study rather than signs of supernatural intention.

Then God became unnecessary at the center of the universe. Copernicus moved Earth from the fixed center. Galileo showed that the heavens were not perfect. Kepler showed that planets moved according to mathematical law. Newton united Earth and heaven under one natural system. Laplace made the point sharper: God was not needed as a working scientific hypothesis.

Then God became unnecessary in the body. Anatomy opened the human form. Physiology explained function. The heart became a pump. Blood became a circulating fluid. Organs, vessels, nerves, tissues, and systems replaced vague ideas of spirits, humors, and mystical forces.

Then God became unnecessary in matter. Chemistry replaced magical transformation with experiment, measurement, oxygen, reactions, gases, conservation, and law.

Then God became unnecessary in Earth's history. Hutton and Lyell revealed deep time. Rocks, fossils, sediments, erosion, uplift, volcanoes, and ancient layers told a story far older than sacred chronology.

Then God became unnecessary in life's apparent design. Darwin showed that life could look designed without being directly designed. Natural selection explained adaptation without foresight, plan, or divine engineering.

Then God became unnecessary in infectious disease. Pasteur, Koch, Semmelweis, Lister, and others helped move sickness from demons, curses, impurity, and divine punishment into microbes, transmission, sanitation, antisepsis, vaccination, antibiotics, and public health.

Then God became less necessary in the universe itself. Einstein changed the meaning of space, time, gravity, and motion. Hubble helped reveal a universe of galaxies, expansion, distance, and deep cosmic history.

Then God became less necessary in parts of the mind. Freud and later psychology made religious belief understandable through fear, father-longing, guilt, comfort, grief, wish, and human need. Neuroscience linked memory, personality, language, emotion, identity, visions, voices, and consciousness to the brain.

This is the great pattern.

God was not defeated by anger.

God became unnecessary through explanation.

The Pattern Is the Argument

This book has followed a simple but powerful historical pattern.

Where human beings lacked knowledge, God was often placed.

Where knowledge grew, God often moved.

This happened again and again.

Before astronomy, God lived in the sky.

Before physics, God moved the heavens.

Before anatomy, God animated the body.

Before chemistry, God or hidden forces transformed matter.

Before geology, God shaped Earth through sacred time and divine catastrophe.

Before evolution, God designed life.

Before germ theory, God punished through disease.

Before psychology, God seemed necessary to explain religious longing.

Before neuroscience, the soul seemed necessary to explain the mind.

In every case, the older supernatural explanation felt natural before evidence-based knowledge existed. Ancient and religious people were not foolish for asking divine questions. They were human. They were trying to live with fear, death, hunger, disease, disaster, mystery, and helplessness.

But understandable does not mean true.

A belief can be emotionally powerful and factually wrong.

A story can be meaningful and scientifically mistaken.

A tradition can be ancient and still fail as an explanation.

The pattern matters because it teaches intellectual caution.

If God has repeatedly been placed in the unknown and repeatedly displaced by natural explanation, then we should be careful before placing God in the mysteries that remain.

Mystery is not evidence.

Ignorance is not proof.

A gap in knowledge is not a doorway through which God automatically enters.

The honest sentence is not, "God did it."

The honest sentence is, "We do not yet know."

That sentence may feel less comforting. But it is more truthful. It keeps the mind open. It allows discovery. It prevents premature certainty. It protects us from turning imagination into doctrine.

What Science Changed

Science did not merely give new answers.

Science changed the way human beings ask questions.

The old religious question often began with agency:

Who caused this?

Which god is angry?

What spirit is present?

What sin produced this suffering?

What divine message is hidden here?

Science asked a different kind of question:

What is happening?

What evidence do we have?

What natural cause can explain it?

Can it be measured?

Can it be tested?

Can it be predicted?

Can the explanation be corrected?

That change transformed civilization.

It gave humanity a way to escape the prison of inherited error. It allowed human beings to correct ancient authority. It allowed observation to challenge doctrine. It allowed instruments to extend the senses. It allowed mathematics to reveal hidden order. It allowed experiments to separate truth from wish.

The telescope changed the sky.

The microscope changed disease.

The balance changed chemistry.

The fossil changed Earth history.

The laboratory changed medicine.

The study of the brain changed the mind.

Each tool, each method, each correction moved reality farther away from supernatural explanation.

Science became powerful not because scientists were always wise or morally pure. They were not. Scientists are human. They can be proud, biased, ambitious, mistaken, and slow to accept correction.

Science became powerful because it developed a method that could correct even scientists.

That is its greatness.

In religion, sacred authority often protects belief from correction.

In science, evidence can defeat authority.

That is why science advanced.

Natural Explanation Proved Itself

Natural explanation did not win only because it sounded logical.

It won because it worked.

Astronomy predicted eclipses.

Physics explained motion.

Chemistry created medicines, materials, and safer methods.

Anatomy guided surgery.

Physiology explained organ function.

Geology read the history of Earth.

Evolution made sense of life's patterns.

Germ theory saved lives.

Neuroscience connected mind, behavior, and brain.

This practical success is one of the strongest arguments in the book.

A prayer may comfort a patient.

But it does not sterilize an instrument.

A sacred story may give meaning to a storm.

But it does not predict the hurricane's path.

A creation story may give identity to a community.

But it does not explain the fossil record.

A belief in divine design may feel satisfying.

But it does not explain antibiotic resistance, genetic similarity, or imperfect anatomy.

A belief in the soul may comfort grief.

But it does not explain why a stroke removes speech or dementia erases memory.

Science changed life because its explanations touched reality.

They could be tested. They could be corrected. They could be used. They could reduce suffering.

That is why natural explanation became so powerful.

It did not merely argue.

It delivered.

The Difference Between Comfort and Explanation

Religion often survives because it offers what science does not try to offer.

Religion gives comfort in grief.

Science explains the cause of death.

Religion gives hope after loss.

Science studies the body and brain.

Religion gives ritual during suffering.

Science studies disease and treatment.

Religion gives a feeling of being watched and loved.

Science studies human need, attachment, fear, and belief.

These are different functions.

We should not pretend that science can replace every emotional role of religion. It cannot. Science does not hold a funeral in the same way religion does. It does not promise reunion after death. It does not say the universe loves us. It does not guarantee justice beyond the grave. It does not tell the grieving mother that her child is safe in heaven.

Science is not a system of comfort.

Science is a system of disciplined truth-seeking.

That is why some people find science emotionally difficult. It does not always give the answer we want. It does not protect us from death. It does not make the universe human-centered. It does not promise that suffering has a hidden purpose.

But there is dignity in this honesty.

False comfort can be costly. If disease is misunderstood as punishment, the sick may be blamed. If disaster is understood as divine judgment, victims may be condemned. If mental illness is seen as possession, patients may be mistreated. If the body is misunderstood, treatment may fail. If prayer is trusted instead of medicine, preventable death may occur.

Comfort matters.

But truth matters too.

And when comfort and truth conflict, mature human beings must not sacrifice truth.

God as Background Belief

As science advanced, God often changed role.

In the oldest explanations, God was the direct cause.

God sent the storm.

God moved the planets.

God shaped each species.

God punished through disease.

God caused fertility or famine.

God gave victory or defeat.

God controlled the sky, Earth, body, and fate.

As scientific explanations grew, this direct God became harder to defend.

So God moved backward.

God became the creator of natural laws.

God became the source of order.

God became the first cause.

God became the reason there is something rather than nothing.

God became the ground of existence.

God became a mystery behind the mysteries.

These ideas are more abstract, and some deserve serious philosophical discussion. Later books in this series will examine them carefully.

But in this book, we should notice the movement itself.

The more science explains, the more God is moved away from direct explanation.

That is not a victory for the old religious view. It is a retreat from it.

A God who once explained lightning but now only explains why there are laws at all has lost explanatory territory. A God who once designed each species but now works through evolution has lost the old design argument. A God who once punished through plague but now is moved behind microbial biology has lost the old disease explanation.

The belief may remain.

But the necessity has weakened.

That is the central point.

Many Scientists Believed in God

This book has been careful about one important fact.

Many great scientists believed in God.

Copernicus was not a modern atheist. Kepler saw mathematical order in religious terms. Newton believed deeply in God. Boyle was religious. Many early scientists worked inside religious cultures. Some saw science as a way to understand divine creation.

This must be stated honestly.

The argument of this book is not that all scientists were atheists.

That would be false.

The argument is stronger:

Even when scientists believed in God, their discoveries often made God less necessary as an explanation.

Newton's personal theology did not prevent his physics from reducing the need for divine management of the heavens. Boyle's religion did not prevent chemistry from moving away from magical views of matter. Pasteur's personal beliefs did not prevent germ theory from replacing supernatural explanations of disease. Darwin's personal journey did not change the power of natural selection as a mechanism for adaptation.

The private belief of a scientist and the public meaning of a scientific explanation are not the same thing.

This distinction protects the argument from unfair exaggeration.

Science made God unnecessary not because every scientist rejected God, but because scientific explanations worked without God.

That is a more careful and more defensible claim.

What This Book Does Not Claim

This book does not claim that science has answered every question.

It has not.

Science does not yet fully explain consciousness. It does not answer every question about the origin of the universe. It does not provide a complete philosophy of meaning. It does not remove death. It does not make grief easy. It does not guarantee justice. It does not tell us how to live in every moral situation.

This book also does not claim that religious people are unintelligent. Many religious people are intelligent, kind, educated, compassionate, and morally serious.

This book does not claim that religion has never done good. Religious communities have cared for the sick, fed the poor, comforted the grieving, built hospitals, created art, inspired courage, and given people strength in suffering.

We should be honest about all of that.

But none of this changes the main argument.

The question in this book has not been, "Has religion ever comforted people?"

Clearly, it has.

The question has not been, “Have religious people ever done good?”

Clearly, they have.

The question has not been, “Can anyone still believe in God after science?”

Clearly, many do.

The question has been more precise:

Is God still necessary as an explanation of reality?

Where science has produced reliable explanations, the answer is no.

The Moral Importance of Natural Explanation

Natural explanation is not merely an intellectual achievement.

It is morally important.

When we stop seeing disease as divine punishment, we treat the sick with more compassion.

When we stop seeing mental illness as possession, we treat the suffering person with more dignity.

When we stop seeing disaster as God’s judgment, we help victims instead of blaming them.

When we stop seeing kings as chosen by heaven, we can question unjust power.

When we stop seeing nature as arranged entirely for us, we may become more humble toward other life.

When we stop seeing human beings as fallen creatures under divine curse, we may see them as vulnerable animals capable of cruelty, kindness, fear, love, reason, and growth.

Science does not automatically make people moral.

Knowledge can be misused. Technology can harm. Scientific institutions can be corrupted. Human beings can turn knowledge into domination.

But accurate explanation gives morality a better foundation than superstition.

If we know the real cause of suffering, we have a better chance of reducing it.

False causes produce false cures.

True causes open the possibility of real help.

That is why this book is not only about God. It is about human responsibility.

If God is not sending the plague, then we must build sanitation.

If God is not causing epilepsy, then we must study the brain.

If God is not punishing the poor through famine, then we must examine agriculture, economics, politics, and justice.

If God is not guiding every disaster, then we must prepare, protect, rescue, and rebuild.

Natural explanation places responsibility back in human hands.

That can be frightening.

It is also liberating.

Wonder After God

Some people fear that a world without divine explanation becomes empty.

They imagine that if God does not move the stars, then the stars lose beauty. If God does not design life, then life loses value. If God does not send the rain, then rain becomes only water. If God does not create the soul, then love becomes only chemistry. If God does not guide history, then life becomes meaningless.

This fear is understandable.

But it is mistaken.

Understanding does not destroy wonder.

An eclipse is still magnificent when we know its cause.

A star is still beautiful when we know it is a sphere of hot plasma powered by nuclear fusion.

A flower is still delicate when we know it evolved through natural selection.

A child is still precious when we know the biology of development.

The brain is still astonishing when we know that memory, emotion, language, and imagination arise from neural activity.

Truth deepens wonder because it shows us what is actually there.

A supernatural story may make the world feel enchanted, but science reveals a deeper enchantment: reality itself is more strange, subtle, vast, and beautiful than ancient imagination knew.

The universe does not need to be divine to be magnificent.

Life does not need to be designed to be precious.

The mind does not need to be a soul to be profound.

Morality does not need heaven to matter.

Human love does not need eternity to be real.

The Shrinking Space for God

At the beginning of human history, God had a large territory.

God lived in the storm, the harvest, the womb, the grave, the king, the plague, the battlefield, the stars, the earthquake, the dream, the fever, the animal, the river, the Sun, and the unknown.

Today, that territory is much smaller.

Thunder belongs to atmospheric physics.

Eclipses belong to astronomy.

Planetary motion belongs to celestial mechanics.

The body belongs to anatomy and physiology.

Matter belongs to chemistry.

Earth history belongs to geology.

Life's diversity belongs to evolution.

Infectious disease belongs to microbiology and medicine.

The expanding universe belongs to cosmology.

Memory and personality belong increasingly to neuroscience.

God has not disappeared from human culture. God has not disappeared from poetry, ritual, music, grief, politics, identity, fear, or hope. God remains emotionally powerful for billions.

But as an explanation of nature, God has shrunk.

That is the honest conclusion of this book.

This shrinking space does not prove every possible form of atheism by itself. It does not answer every philosophical question. It does not settle all debates about first cause, consciousness, morality, evil, prayer, miracles, or scripture.

Those questions will come in later books.

But the historical pattern is clear.

God was humanity's first answer to the unknown.

Science has shown, again and again, that the first answer was not the best answer.

From Historical Retreat to Cosmic Question

This book has given the historical map.

It has shown how science gradually replaced divine explanation in one field after another. The movement began with the visible world around us and reached outward into the heavens and inward into the mind.

But one great question now stands before us.

Even if God is not needed to explain thunder, disease, evolution, geology, or planetary motion, what about the universe itself?

Does the universe need a creator?

Does the Big Bang prove God?

Does the existence of physical law require a divine lawgiver?

Does fine-tuning point to design?

Does the mystery of existence lead back to a first cause?

These are serious questions. They should not be dismissed casually. They deserve careful examination, because for many thoughtful believers, the strongest remaining argument for God is no longer thunder, disease, or design in living bodies. It is the universe itself.

That is why the next book in this series turns to cosmology.

Book 3 will examine the universe without a creator.

It will explore relativity, Big Bang cosmology, first-cause arguments, fine-tuning, the anthropic principle, multiverse ideas, quantum cosmology, and Stephen Hawking's challenge to the need for God as creator.

For now, the conclusion of Book 2 is this:

Science has not removed every mystery.

But wherever science has produced reliable explanations, God has become unnecessary.

That is not a small fact.

It is one of the most important facts in the modern understanding of reality.

Final Thought

Human beings created gods because they needed answers.

Science did not take away the need for answers. It gave us a better way to search for them.

The older world looked at thunder and heard divine anger. Science heard electricity.

The older world looked at fever and saw punishment. Science found infection.

The older world looked at life and saw direct design. Science found evolution.

The older world looked at the heavens and saw sacred order. Science found planets, stars, galaxies, gravity, space-time, and expansion.

The older world looked at the mind and saw the soul. Science found the brain.

This is not a loss of wonder.

It is the growth of understanding.

God was once the explanation for almost everything.

Then humanity learned to look more carefully.

And slowly, patiently, courageously, reality began to explain itself.

Sources

Nicolaus Copernicus, *On the Revolutions of the Heavenly Spheres*

Galileo Galilei, *Sidereus Nuncius*

Johannes Kepler, *Astronomia Nova*

Isaac Newton, *Mathematical Principles of Natural Philosophy*

Pierre-Simon Laplace, *Celestial Mechanics*

Andreas Vesalius, *De Humani Corporis Fabrica*

William Harvey, *On the Motion of the Heart and Blood*

Robert Boyle, *The Sceptical Chymist*

Antoine Lavoisier, *Elements of Chemistry*

James Hutton, *Theory of the Earth*

Charles Lyell, *Principles of Geology*

Charles Darwin, *On the Origin of Species*

Louis Pasteur, scientific writings on fermentation, spontaneous generation, and germ theory

Robert Koch, scientific writings on bacteriology

Albert Einstein, *Relativity: The Special and the General Theory*

Edwin Hubble, writings on galaxies and cosmic expansion

Sigmund Freud, *The Future of an Illusion*

William James, *The Varieties of Religious Experience*

Thomas Kuhn, *The Structure of Scientific Revolutions*

John Hedley Brooke, *Science and Religion: Some Historical Perspectives*

Peter Harrison, *The Territories of Science and Religion*

Jacob Bronowski, *The Ascent of Man*

Carl Sagan, *Cosmos*

Carl Sagan, *The Demon-Haunted World*

Daniel Dennett, *Breaking the Spell*

Sean Carroll, *The Big Picture*

Stanford Encyclopedia of Philosophy, "Religion and Science"

Encyclopedia Britannica, "Scientific Revolution"

National Academy of Sciences, *Science, Evolution, and Creationism*

Appendix: Scientists, Scholars, and Books That Changed the Human Understanding of Reality

Why This Appendix Is Included

The argument of this book did not arise from one discovery, one scientist, one philosopher, or one famous sentence. It emerged from a long historical movement in which human beings slowly learned to explain nature through observation, measurement, experiment, mathematics, and evidence.

Across many centuries, thinkers from different backgrounds helped move human understanding away from supernatural explanation and toward natural explanation. Some were deeply religious. Some were skeptical. Some avoided theological questions altogether. Some studied the sky. Some studied the body. Some studied matter, disease, Earth, life, the mind, or the universe.

They did not all share one philosophy. They did not all reach the same conclusions about God. Many did not intend to weaken religion. Yet their work changed the human understanding of reality in a profound way.

They showed, field by field, that nature could increasingly explain itself.

This appendix is not a complete history of science. It is a selected guide to important figures and books that support the central theme of this volume: as scientific knowledge expanded, God became less necessary as an explanation of the natural world.

Astronomy and the Scientific Revolution

Nicolaus Copernicus

Major work: *On the Revolutions of the Heavenly Spheres*

Copernicus changed the human position in the cosmos. By placing the Sun near the center of the planetary system and making Earth a moving planet, he weakened the older Earth-centered worldview.

His work did not disprove God. But it challenged the comforting belief that humanity lived at the fixed center of creation. It began one of the greatest intellectual displacements in history: the movement of human beings away from cosmic centrality.

Copernicus matters in this book because he helped begin the long transition from sacred sky to physical universe.

Galileo Galilei

Major works:
Sidereus Nuncius
Dialogue Concerning the Two Chief World Systems

Galileo helped bring the heavens into the world of evidence. With the telescope, he observed mountains on the Moon, moons around Jupiter, phases of Venus, sunspots, and many more stars than the naked eye could see.

These discoveries weakened the old belief that the heavens were perfect, pure, and fundamentally different from Earth. They also showed that observation could challenge inherited religious and philosophical authority.

Galileo did not destroy religion. But he helped destroy the old sacred astronomy.

Johannes Kepler

Major works:
Astronomia Nova
Harmonices Mundi
Epitome of Copernican Astronomy

Kepler showed that the planets move according to mathematical laws. He replaced the ancient dream of perfect circular motion with elliptical orbits and demonstrated that the sky was not a field of mystical wandering, but a lawful natural system.

Kepler himself was religious and often saw mathematical order in religious terms. Yet his science made celestial motion less dependent on divine or mystical explanation.

He helped move the sky from omen to orbit.

Isaac Newton

Major works:
Mathematical Principles of Natural Philosophy
Opticks

Newton brought Earth and heaven under one law. His laws of motion and universal gravitation showed that the same principles could explain falling objects on Earth and the motion of the Moon and planets in the sky.

Newton believed deeply in God. Yet the scientific meaning of his work was revolutionary: the heavens could be understood through natural law rather than direct divine management.

His work made the universe more lawful, unified, mathematical, and predictable.

Pierre-Simon Laplace

Major works:
Celestial Mechanics
A Philosophical Essay on Probabilities

Laplace represents a decisive stage in the retreat of God as a scientific explanation. His famous reply that he had no need of the God hypothesis captures a central idea of this book: if natural law explains the evidence, God is not needed as part of the scientific model.

Laplace did not settle every philosophical question about God, existence, or the origin of natural law. But he made clear that celestial mechanics could proceed without invoking divine action.

In his work, the heavens no longer required God as a working hypothesis.

Anatomy, Physiology, and Medicine

Galen

Major works:
On the Natural Faculties
Other medical and anatomical writings

Galen shaped medical thought for centuries. He was a serious and influential physician whose writings preserved important observations. But his long authority also illustrates one of the dangers of intellectual tradition: when a respected figure becomes almost untouchable, even errors can survive for generations.

Galen matters in this book because later anatomy and physiology had to move beyond inherited authority and return to the body itself.

His legacy shows both the value and the danger of ancient knowledge.

Andreas Vesalius

Major work: *De Humani Corporis Fabrica*

Vesalius helped move anatomy away from blind dependence on ancient texts and toward direct observation of the human body. His dissections and illustrations corrected many inherited errors and changed the study of medicine.

His work showed that the body itself must be the authority.

Vesalius helped move the body from sacred mystery to natural structure.

William Harvey

Major work: *On the Motion of the Heart and Blood*

Harvey showed that blood circulates through the body and that the heart functions as a pump. His work helped transform physiology through observation, measurement, experiment, and reasoning.

The heart remained symbolically powerful, but medically it became a physical organ with a natural function.

Harvey helped replace mystery with physiology.

Ignaz Semmelweis

Major work: *The Etiology, Concept, and Prophylaxis of Childbed Fever*

Semmelweis showed that handwashing could dramatically reduce deaths from childbirth fever. His evidence was resisted, partly because it implied that physicians themselves were carrying deadly material to their patients.

His story remains one of the painful lessons in medical history. Evidence can be resisted when it humiliates authority.

Semmelweis helped prepare the way for germ theory and modern infection control.

Louis Pasteur

Major works:
Scientific writings on fermentation, spontaneous generation, microbes, and disease

Pasteur helped reveal the microbial world. His work on fermentation, contamination, spontaneous generation, pasteurization, and disease showed that invisible organisms could produce visible effects.

He weakened older ideas of disease as curse, impurity, divine punishment, or mysterious corruption.

Pasteur helped make disease natural.

Robert Koch

Major works:
Scientific writings on anthrax, tuberculosis, cholera, and bacteriology

Koch helped connect specific microbes with specific diseases. His work gave germ theory greater precision and practical force.

Disease was no longer a vague curse or spiritual condition. It could be investigated, cultured, identified, and linked to particular organisms.

Koch helped move medicine from fear to evidence.

Joseph Lister

Major works:
Writings on antiseptic surgery

Lister applied germ theory to surgery. By using antiseptic methods to reduce contamination, he helped make surgery safer and more scientific.

His work showed the life-saving power of correct explanation. If wound infection is microbial, then prevention requires cleanliness, antisepsis, and later aseptic technique, not ritual or prayer alone.

Lister helped turn surgery from a desperate gamble into a more disciplined medical practice.

Chemistry and Matter

Robert Boyle

Major work: *The Sceptical Chymist*

Boyle helped move chemistry away from older elemental and alchemical ideas toward experiment and careful reasoning. He challenged inherited theories and insisted that matter should be studied through evidence.

Boyle was personally religious, but his scientific work helped weaken magical explanations of substance and transformation.

He helped make matter less mystical and more measurable.

Antoine Lavoisier

Major works:
Elements of Chemistry
Writings on combustion and oxygen

Lavoisier transformed chemistry through careful measurement, clear language, and the oxygen theory of combustion. He helped replace the mistaken phlogiston theory and emphasized conservation of matter in chemical reactions.

His work showed that burning, rusting, and chemical change could be explained naturally.

Matter no longer needed hidden fire spirits, occult qualities, or magical essences.

Geology and Deep Time

James Hutton

Major work: *Theory of the Earth*

Hutton helped introduce the idea of deep time. He saw Earth as shaped by natural processes such as erosion, sedimentation, heat, uplift, and renewal over immense periods.

His work weakened sacred chronologies and young-Earth thinking.

Hutton helped show that Earth had a natural history far older than human memory, religious genealogy, or sacred timeline.

Charles Lyell

Major work: *Principles of Geology*

Lyell argued that present natural processes could explain much of Earth's past when given enough time. His work popularized and strengthened the idea of gradual geological change over vast periods.

Lyell influenced Darwin and helped provide the timescale that evolution required.

He helped move Earth history from sacred story to geological evidence.

Evolution and Life

Charles Darwin

Major works:
On the Origin of Species
The Descent of Man

Darwin gave science one of its most powerful natural explanations: evolution by natural selection. He showed how life could look designed without being directly designed.

His work weakened special creation and the traditional design argument.

Darwin did not answer every question about life, and modern genetics later strengthened evolution greatly. But his central achievement remains one of the greatest turning points in the retreat of divine explanation.

He made biological design understandable without a divine designer.

Alfred Russel Wallace

Major works:
Writings on natural selection and biogeography

Wallace independently developed the idea of natural selection and contributed significantly to biogeography. His work strengthened the evidence that species are shaped by environment, geography, variation, and history.

Wallace's own religious and spiritual views were complex, but his scientific work supported a natural explanation for adaptation and species change.

He belongs in this appendix because natural selection was not merely one man's private insight. It was an idea emerging from evidence.

Physics and Cosmology

Albert Einstein

Major works:
Relativity: The Special and the General Theory
Scientific papers on special relativity, general relativity, and cosmology

Einstein changed the meaning of space, time, gravity, and motion. His work showed that the universe is stranger, deeper, and more lawful than common sense suggests.

General relativity made gravity part of the structure of space-time and prepared the way for modern cosmology.

Einstein did not remove all mystery from the universe, but he helped make the cosmos a subject of physical law rather than sacred architecture.

Edwin Hubble

Major works:
Scientific papers on galaxies, Cepheid variables, and cosmic expansion

Hubble helped show that the Milky Way is not the whole universe and that distant galaxies are moving away from us. His work enlarged the universe dramatically and helped establish the idea of cosmic expansion.

Hubble's observations made the old human-centered cosmos even less defensible.

The universe became vast, dynamic, and filled with galaxies beyond ancient imagination.

Georges Lemaître

Major works:
Scientific writings on the expanding universe and the "primeval atom"

Lemaître, a physicist and Catholic priest, proposed an expanding universe model and the idea that the universe developed from an earlier compact state. His life is especially important because it shows that religious identity and scientific explanation are not always simple enemies.

Yet his scientific work also helped move cosmology into natural explanation.

The universe could be studied through physics, not simply through theology.

Psychology, Mind, and Religion

Sigmund Freud

Major works:
The Future of an Illusion
Totem and Taboo
Civilization and Its Discontents

Freud treated religion as a psychological phenomenon. He argued that God-belief is connected to wish fulfillment, father-longing, helplessness, guilt, fear, and the need for protection.

Freud's ideas should be read with care. He was historically important, but not always correct in detail. Still, his work helped move religion from heaven into the human mind.

He asked a powerful question: What human needs does God satisfy?

William James

Major work: *The Varieties of Religious Experience*

William James studied religious experience with seriousness and sympathy. He did not simply mock believers. He examined conversion, mysticism, prayer, emotion, and personal religious life as human experience.

James is useful because he shows that religious experiences can be psychologically real and personally powerful, even when their interpretation remains open to question.

He helps us separate the reality of an experience from the truth of the explanation attached to it.

Daniel Dennett

Major work: *Breaking the Spell*

Dennett argues that religion should be studied as a natural phenomenon. He examines belief, ritual, culture, evolution, and social function without assuming supernatural truth.

His work supports the idea that religion can be explained through human minds, cultures, and evolutionary processes.

For this series, Dennett is useful because he treats religion not only as doctrine, but as a human phenomenon that can be studied.

Pascal Boyer

Major work: *Religion Explained*

Boyer uses cognitive science and anthropology to explain why religious ideas are natural for human minds. He shows how concepts of gods, spirits, ancestors, and invisible agents fit human cognitive tendencies.

His work supports the argument that belief in gods may arise naturally from how the human mind works.

Boyer is especially useful for understanding why supernatural ideas are so memorable and culturally durable.

Scott Atran

Major work: *In Gods We Trust*

Atran examines religion through cognitive science, anthropology, and evolutionary psychology. His work explores why religious beliefs are widespread, emotionally powerful, and socially persistent.

He is useful for understanding religion as a natural human phenomenon rather than as direct evidence of divine revelation.

Michael Shermer

Major work: *The Believing Brain*

Shermer argues that the human brain tends to find patterns and then attach beliefs to those patterns. His work is useful for understanding why people may see agency, intention, and meaning in random or natural events.

This supports one of the central themes of Book 1 and Book 2: human beings are natural believers because the mind is built to seek pattern and agency.

Neuroscience and the Natural Mind

Antonio Damasio

Major works:
Descartes' Error
Self Comes to Mind

Damasio's work connects emotion, body, brain, and consciousness. He helps show that the mind is embodied, not floating separately from the physical organism.

His work supports the argument that the self is deeply tied to brain and body.

Oliver Sacks

Major works:
The Man Who Mistook His Wife for a Hat
An Anthropologist on Mars

Sacks wrote with unusual humanity about patients with neurological disorders. His cases show how changes in the brain can alter perception, identity, memory, movement, and personality.

His work helps readers understand that the mind is not separate from the brain in any simple way.

He also shows that a natural view of the mind need not be cold. It can be deeply humane.

V. S. Ramachandran

Major work: *Phantoms in the Brain*

Ramachandran explores neurological cases that reveal the brain's role in body image, perception, selfhood, and unusual experiences. His writing helps show how strange and powerful brain-based experience can be.

His work is useful for understanding why visions, altered states, and self-experience may have natural explanations.

Michael Gazzaniga

Major works:
The Ethical Brain
Who's in Charge?

Gazzaniga's work on split-brain research and cognitive neuroscience is important for understanding how the self is constructed by brain processes. His work challenges simple ideas of a unified, immaterial soul.

He is useful for readers who want to understand why the self is not as simple as it feels from the inside.

Patricia Churchland

Major works:
Neurophilosophy
Braintrust

Churchland connects neuroscience with philosophy, morality, and the mind. Her work supports a naturalistic understanding of consciousness, behavior, and moral life.

She is especially useful for later books in this series, but her work also supports the basic argument that mind depends on brain.

Philosophy, History, and the Meaning of Science

David Hume

Major works:
The Natural History of Religion

Dialogues Concerning Natural Religion
An Enquiry Concerning Human Understanding

Hume remains one of the most important skeptical philosophers. He examined religion, miracles, causation, design, and human belief with unusual clarity.

His work is useful across the whole series. For this book, he helps frame the problem of religious explanation and the weakness of arguments based on ignorance, testimony, and apparent design.

Thomas Kuhn

Major work: *The Structure of Scientific Revolutions*

Kuhn examined how scientific revolutions occur. His concept of paradigm shifts helps explain why major changes such as the Copernican revolution, Newtonian mechanics, Darwinian evolution, and modern physics were not merely additions of facts, but transformations of worldview.

Kuhn is important for understanding why old explanations resist replacement.

John Hedley Brooke

Major work: *Science and Religion: Some Historical Perspectives*

Brooke is especially important because he avoids simplistic conflict stories. He shows that the relationship between science and religion has been complex, varied, and historically rich.

His work helps keep this book honest. The argument is not that science and religion have always been simple enemies, but that scientific explanation has repeatedly reduced the need for divine explanation.

Peter Harrison

Major work: *The Territories of Science and Religion*

Harrison examines how the categories of "science" and "religion" developed historically. His work helps prevent careless anachronism and reminds us that modern terms cannot always be imposed simply on earlier periods.

This is useful for maintaining historical seriousness and accuracy.

David Wootton

Major work: *The Invention of Science*

Wootton presents the rise of modern science as one of the great transformations in human thought. His work is useful for understanding how experiment, discovery, evidence, and new methods changed the intellectual world.

He helps readers see science not merely as a collection of facts, but as a new way of knowing.

Alexandre Koyré

Major work: *From the Closed World to the Infinite Universe*

Koyré examined the transformation from the older closed cosmos to the modern idea of an immense, law-governed universe. His work is especially useful for understanding Copernicus, Galileo, Kepler, Newton, and the changing meaning of the heavens.

He helps show that modern science did not merely change astronomy. It changed the human picture of reality.

Jacob Bronowski

Major work: *The Ascent of Man*

Bronowski presents science as a deeply human achievement, rooted in imagination, courage, and moral responsibility. His work is useful because it shows science not as a cold machine, but as one of humanity's noblest forms of inquiry.

His tone is especially valuable for this series: humane, serious, and filled with wonder.

Carl Sagan

Major works:
Cosmos
The Demon-Haunted World

Sagan combined scientific wonder with skepticism. He showed that science can inspire awe without supernatural belief. He also warned against superstition, pseudoscience, and false certainty.

Sagan is valuable because he demonstrates that a natural view of reality can be beautiful, ethical, and deeply moving.

Sean Carroll

Major works:
The Big Picture
From Eternity to Here

Carroll presents a modern naturalistic view of reality, including physics, cosmology, time, complexity, life, and meaning. His work is especially useful for later books in this series, but it also supports the broader argument that reality can be understood without invoking God.

He helps connect modern science with a thoughtful naturalistic worldview.

Other Important Works in Modern Religious Criticism

The following books are important in the modern debate about religion, atheism, and reason. They differ in tone, style, and emphasis, but each has contributed to public discussion about faith, evidence, skepticism, and the rational defense or criticism of belief.

Richard Dawkins, *The God Delusion*

Christopher Hitchens, *God Is Not Great*

Sam Harris, *The End of Faith*

A. C. Grayling, *The God Argument*

Bertrand Russell, *Why I Am Not a Christian*

J. L. Mackie, *The Miracle of Theism*

Victor Stenger, *God: The Failed Hypothesis*

These works are included because they represent major modern critiques of religious belief. Readers may agree or disagree with their style or conclusions, but they are part of the wider intellectual conversation about whether belief in God remains rationally justified.

Suggested Reading Path

Readers who want a broad understanding of science, religion, skepticism, and natural explanation may begin with the following works:

Carl Sagan, *Cosmos*

Carl Sagan, *The Demon-Haunted World*

Jacob Bronowski, *The Ascent of Man*

John Hedley Brooke, *Science and Religion: Some Historical Perspectives*

Peter Harrison, *The Territories of Science and Religion*

David Wootton, *The Invention of Science*

Thomas Kuhn, *The Structure of Scientific Revolutions*

Sean Carroll, *The Big Picture*

Daniel Dennett, *Breaking the Spell*

Pascal Boyer, *Religion Explained*

Readers who want to explore the scientific foundations more directly may continue with:

Copernicus, *On the Revolutions of the Heavenly Spheres*

Galileo, *Sidereus Nuncius*

Kepler, *Astronomia Nova*

Newton, *Mathematical Principles of Natural Philosophy*

Laplace, *Celestial Mechanics*

Vesalius, *De Humani Corporis Fabrica*

Harvey, *On the Motion of the Heart and Blood*

Boyle, *The Sceptical Chymist*

Lavoisier, *Elements of Chemistry*

Hutton, *Theory of the Earth*

Lyell, *Principles of Geology*

Darwin, *On the Origin of Species*

Einstein, *Relativity: The Special and the General Theory*

Closing Note

The figures and books listed here do not speak with one voice. They do not represent one party, one ideology, or one simple conflict between science and religion. Many of the scientists named here believed in God. Some were skeptical. Some were religious in ways that do not fit neatly into modern categories.

But taken together, their work changed the world.

They showed that the sky could be explained without divine messages.

The planets without angelic movers.

The body without mystical forces.

Matter without magic.

Earth without sacred chronology.

Life without special creation.

Disease without demons.

The universe without a small human-centered design.

The mind without an obvious supernatural soul.

This does not mean every mystery has vanished. It means the old habit of using God as the first explanation has lost its authority.

Science has not answered everything.

But it has answered enough to change the question forever.

No God is required where nature explains itself.

Thank You and Please Review the Book

Thank you for taking the time to read this book. I sincerely hope it has encouraged you to think more deeply, question more honestly, and look at religion, science, and human belief with greater clarity.

My goal in writing the **No God Required Series** is simple: to present serious, science-based, and easy-to-understand ideas about God, religion, nature, morality, and human meaning. These books are written for readers who value evidence, reason, intellectual honesty, and the courage to ask difficult questions.

If you found this book thoughtful, helpful, or worth reading, I would be truly grateful if you could leave a brief and honest review. Even a few sincere lines can make a meaningful difference. Your review helps other readers discover this series and encourages more open, rational, and evidence-based discussion about some of life's deepest questions.

Thank you again for your time, your trust, and your support.

Warm regards,
Dr. Prabhat Das